Handbook of
Ventilation
for Contaminant
Control

Handbook of
Ventilation
for Contaminant
Control

Second Edition

Henry J. McDermott

BUTTERWORTH PUBLISHERS
Boston • London
Sydney • Wellington • Durban • Toronto
An Ann Arbor Science Book

Ann Arbor Science is an imprint of Butterworth Publishers.

Copyright © 1985 by Butterworth Publishers.
All rights reserved.

Library of Congress Cataloging in Publication Data

McDermott, Henry J.
  Handbook of ventilation for contaminant control.

  "An Ann Arbor Science book."
  Includes bibliographies.
  1. Ventilation—Handbooks, manuals, etc.
  2. Decontamination (from gases, chemicals, etc.)—
  Handbooks, manuals, etc.   I. Title.
  TH7656.M3   1985   697.9'2   84–29321
  ISBN 0–250–40641–1

Butterworth Publishers
80 Montvale Avenue
Stoneham, MA 02180

10 9 8 7 6 5 4 3 2

Printed in the United States of America

Contents

List of Tables

Preface

Ventilation is a key method for reducing employee exposures to airborne contaminants resulting from industrial operations. Ventilation can be used either to dilute contaminants to safe levels or to capture and remove them at their sources before they pollute the working environment. Occupational Safety and Health Act (OSHA) standards setting legal limits on employee exposures focus attention on the need to identify and reduce hazardous exposures.

Despite OSHA, many plants still lack ventilation in areas where it is needed, and too many systems that are installed do not work properly. Designing a ventilation system that will work is not a difficult task if a few basic airflow principles are followed. Hoods for capturing or containing contaminants must be selected with the industrial process and contaminants in mind. Ducts, air cleaner, and fan must work together to draw the needed airflow through each hood in the system.

In most plants, exhaust ventilation is the responsibility of the safety department and plant engineering staff. Often the safety department identifies the potentially harmful exposures and the plant engineer designs the system hardware and has it installed. Whether ventilation is used where it is needed and whether it works properly to provide adequate protection depends on the knowledge and skill of the plant safety and engineering staffs.

The first edition of this book was written to fill a gap, since there was no practical and comprehensive text covering ventilation in a manner that safety professionals, industrial hygienists, plant engineers, and others could understand and use on their own. An excellent design manual is available, but it does not contain enough background and theory to be self-explanatory. Professional development courses are offered by universities and professional associations; they typically emphasize the design of new systems rather than methods for expanding or upgrading existing systems.

This revised edition contains two new chapters (Chapter 2 on indoor air pollution and ventilation and Chapter 7 on air cleaners). In addition, Chapter 8 (System Design) has been extensively rewritten, since feedback from readers indicated that some found the design examples in the previous edition difficult to follow. For this reason the velocity pressure calculation sheet has been streamlined, the step-by-step explanations have been expanded, and another sample calculation has been added. This edition also includes a blank calculation sheet to be photocopied for the reader's use.

The text is a personal view of ventilation; the topics and examples reflect what I find most important. The helpful suggestions from readers of the first edition have helped me to improve this revised edition.

H. J. McDermott

CHAPTER 1

Introduction

Ventilation is an important method of reducing exposures to airborne contaminants. It also serves other purposes, such as preventing the accumulation of flammable or explosive concentrations of gases, vapors, or dusts in the workplace. The proper design and use of ventilation systems require an understanding of the principles governing their operation and insight into practical solutions to exposure problems.

Not too long ago any discussion of workplace exposures to airborne contaminants focused only on industrial operations. Recently, however, the problem of employee exposures to indoor air pollutants in office and commercial buildings has emerged as an important issue in worker health and safety.

The goal of this book is to provide information on how to use ventilation effectively and economically to solve employee exposure problems. The following is a chapter-by-chapter summary of how the book organizes this information:

- Chapter 1 differentiates between local exhaust ventilation and dilution ventilation and presents the technical basis for dilution ventilation in industrial settings.
- Chapter 2 discusses indoor air pollution in office and commercial environments and ventilation as cause and solution in indoor air quality problems.
- Chapter 3 begins the focus on local exhaust ventilation and describes OSHA standards relating to ventilation.
- Chapter 4 discusses how to determine whether an overexposure exists, in order to decide whether ventilation or another control is needed.
- Chapter 5 explains how a local exhaust ventilation system works to reduce contaminant levels.
- Chapter 6 describes the most important part of the system, the hoods. If the wrong hoods are selected, the system may never reduce exposures sufficiently to meet OSHA standards.
- Chapter 7 introduces air pollution control equipment used in ventilation systems.

- Chapters 8 and 9 explain how to design the ducts and select the fan so that the correct amount of air is drawn in through the hoods.
- Chapter 10 discusses some special techniques for ventilation systems handling highly toxic contaminants, since standard systems may not reduce airborne levels enough to bring exposures to highly toxic materials to a safe level.
- Chapter 11 discusses how to provide ventilation at the lowest cost by balancing capital and operating costs.
- Chapter 12 explains how to test ventilation systems. More and more OSHA standards require periodic testing.
- Chapter 13 presents ways to solve problems in existing ventilation systems.

DILUTION VERSUS LOCAL EXHAUST

There are two major types of industrial ventilation: dilution, or general ventilation, and local exhaust.

Dilution Ventilation

Dilution occurs when contaminants released into the workroom mix with air flowing through the room (Figure 1.1). Either natural or mechanically-induced air movement can be used to dilute contaminants. Dilution ventilation is used in situations meeting these criteria:[1]

- Small quantities of contaminants released into the workroom at fairly uniform rates.

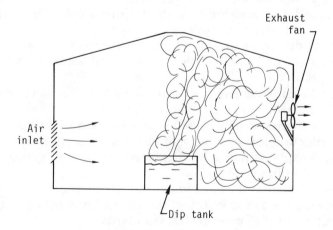

Figure 1.1 Dilution ventilation reduces contaminant levels as fresh air mixes with the contaminants in the workroom air.

- Sufficient distance from the worker (or source of ignition for fire/ explosion hazards) to the contaminant source to allow dilution to safe levels.
- Contaminants of low toxicity or fire hazard.
- No air-cleaning device needed to collect contaminants before the exhaust air is discharged into the community environment.
- No corrosion or other problems from the diluted contaminants in the workroom air.

The major disadvantages of dilution ventilation are that large volumes of dilution air may be needed and that employee exposures are difficult to control near the contaminant source where dilution has not yet occurred.

Dilution ventilation is also called general ventilation. However, in many industrial plants the overall heating and cooling air system is referred to as the general ventilation system. Therefore, to avoid confusion, the term *dilution* will be used for contaminant control systems. The design of dilution ventilation systems is discussed later in this chapter.

Local Exhaust Ventilation

Local exhaust systems capture or contain contaminants at their source before they escape into the workroom environment. A typical system consists of ducts, one or more hoods, an air cleaner if needed, and a fan (Figure 1.2). The big advantage of local exhaust systems is that they remove contaminants rather than just dilute them. Even with local exhaust some airborne contaminants may still be in the workroom air due to uncontrolled sources or less than 100% collection efficiency at the hoods. A second major advantage of local exhaust is that these systems require less airflow than do dilution ventilation systems in the same applications. The total airflow is important for plants that are heated or cooled since heating and air conditioning costs are an important operating expense. These cost factors are discussed in Chapter 11.

Local exhaust systems may be more difficult to design than dilution systems. The hoods or pick-up points must be properly shaped and located to control contaminants, and the fan and ducts must be designed to draw the correct amount of air in through each hood.

Purpose of this Book

The goal of this book is to help you use ventilation, especially local exhaust ventilation, effectively. If you understand how local exhaust systems work and follow a few basic airflow principles during design,

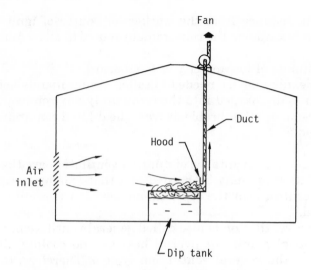

Figure 1.2   Local exhaust ventilation captures or contains contaminants at their source before they disperse in the work environment.

your ventilation system will provide the protection to employees that you intend. Understanding the way in which the system functions will also help you to diagnose and correct problems in existing ventilation systems.

One idea repeated several times in this text is that the hoods are the most important part of the local exhaust system. There are three different types of hoods: those that enclose the contaminant source; those that are positioned to catch a stream of contaminants thrown out by a source or released in a given direction (such as a hot dip tank); and those that reach outside the hood to capture contaminants. The welding hood shown in Figure 1.3 is a popular type of capturing hood. Chapter 6 describes the different hood types and their use, but a vital phase of any ventilation project is deciding which type of hood is best for each contaminant source and then choosing the airflow required for each hood to work properly. If the original hood selection is wrong, no matter how well the ducts and fan are designed, the system will probably never reduce airborne contaminants sufficiently. Hood selection is based on the characteristics of the contaminants, the way contaminants disperse when released, and the physical layout of the plant.

USES OF VENTILATION

Control of contaminants in the workroom air for health protection or fire/explosion prevention is not the only use for ventilation. Other uses include comfort, material reuse, and environmental protection.

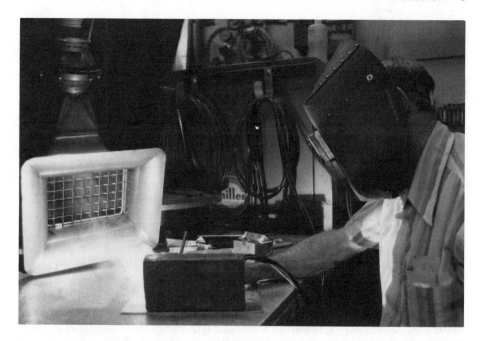

Figure 1.3 Welding hoods have sufficient airflow to reach outside the hood to capture welding fumes.

Comfort

Heat, odors, and tobacco smoke can be removed from working or living areas by exhaust ventilation. Air supply systems furnish heated or cooled air depending on the season. Heat can, of course, be a health hazard in extreme exposures, and then heat control is more than just a comfort consideration.[2] Dilution, or general ventilation, is most often used for comfort ventilation to control temperature, humidity, and radiant heat load and to provide fresh air circulation without drafts.

Material Control and Reuse

Local exhaust ventilation can be used to conserve reusable materials or those that cause housekeeping problems. It is advantageous to install ventilation if a reusable material that is released from a process can be collected by a ventilation system and then removed from the airstream for recycling using an air cleaner at a lower cost than replacing the material. Other waste materials cause a housekeeping problem rather than a health concern. These materials can also be collected in a ventilation system and removed for disposal with an air cleaner. Sawdust from woodworking shops is a good example. Although very fine sawdust can remain airborne and presents a fire or health hazard, most of the material collected in the local exhaust system is too large to remain airborne and so piles up on the floor around the equipment. It is often more efficient to collect this material in a local exhaust system than to sweep it up.

Community Environmental Protection

Environmental protection is linked to material reuse in that similar ventilation systems are used to control the contaminants produced by the industrial process, and air cleaners are used to remove the contaminants before the air is discharged to the community environment. For environmental protection systems, however, the aim is to remove materials so that the residual in the discharged air meets air pollution standards; the material collected in the air cleaner may have no reuse value.

Some environmental regulations set limits on the amounts of pollutants that can be discharged from specific industrial operations. For example, U.S. Environmental Protection Agency standards[3] govern the discharge of "Hazardous Air Pollutants," including asbestos and mercury, to the atmosphere. To meet these standards, a local exhaust system with the proper air-cleaning device may be needed.

Many of the ventilation system features for reducing employee exposures also apply to systems for material recycling or environmental conservation.

OTHER EXPOSURE CONTROL TECHNIQUES

Ventilation is only one way to reduce employee exposures. An employee's exposure to an airborne substance is related to the amount of contaminants in the air and the time during which the employee breathes the concentration. Any factors that interrupt the exposure pattern by reducing the amount of contaminant in the employee's breathing zone or the amount of time that the employee spends in the area will reduce his overall exposure. To lower exposures, review the contaminant source, the path it travels to the worker, and the employee's work pattern and use of protective equipment.[4]

Source Control

In addition to local exhaust ventilation, here are some ways to reduce the amount of contaminants released into the workroom air:

- Preventive maintenance to repair leaks or control other factors that increase emissions.
- Substitution of less toxic materials for those already in use. Benzene and carbon tetrachloride are examples of solvents that were once in wide use but have been replaced with less toxic solvents where feasible. The substitution of artificial abrasives for sand in abrasive blasting cleaning operations to reduce exposure to free silica is another illustration.
- A change in the process to reduce the amount of contaminant released. For example, the use of a water spray when cutting asbestos-concrete pipe (Figure 1.4) prevents dispersion of asbestos fibers without ventilation. The use of large bulk sacks (Figure 1.5) for dry powders in place of individual smaller paper sacks can reduce the dust emissions that accompany the dumping and disposal of the small sacks.

Exposure Pathway Modifications

The exposure pathway is the route by which the contaminant travels from the source to the worker's breathing zone. When dilution ventilation is used, it affects the exposure pathway because it disperses and removes contaminants. The accumulation around the process or in the workroom is thereby reduced. Other choices include:

- Lengthening the exposure pathway by increasing the distance between source and worker. The dilution rate is influenced by drafts and other air motion, but in almost all cases the contaminant concentration decreases with increased distance from the source.

Figure 1.4   Water spray controls dust and fiber emissions while asbestos-concrete pipe is cut. This is one of several control methods besides ventilation.

Figure 1.5 Bulk sacks with a spout at the bottom to drain
the contents can reduce dust emissions when transferring
dry materials. *(Courtesy B.A.G. Corp.)*

- Interrupting the exposure pathway with physical barriers such as
  doors, curtains, or baffles to impede the movement of contaminated
  air toward workers. Figure 1.6 illustrates the effect of a window fan
  on airborne dust levels near a surface-grinding operation. When the
  fan was blowing across the grinder toward the worker, the airborne
  concentration in the worker's breathing zone was almost five times
  higher than the concentration when the fan was off.[5]
- Isolating the process or the worker. Usually this involves moving one
  or the other to a different room to minimize exposure levels and
  exposure time. This especially applies to workers who are not directly
  involved with the process releasing the contaminants. The degree of

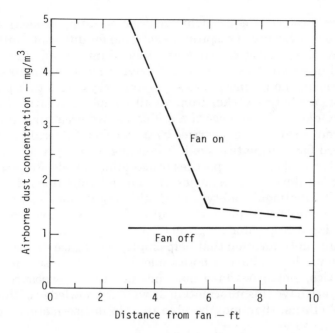

Figure 1.6 Effect of a window fan blowing across a grinding wheel, located 2 feet from the fan, on airborne dust concentrations.

isolation required depends on the toxicity of the contaminant, the amount of contaminant released, and work patterns around the process. Often moving a unit to another room is sufficient, while in other cases a control room supplied with fresh air for the operators may be needed.

Sometimes airborne levels of contaminants can be monitored continuously with a sensor that alarms when concentrations exceed an established level. When an alarm sounds, the workers or supervisors take steps to reduce airborne levels. This approach is useful to detect abnormal operating conditions that cause excessive emissions.

Other Steps to Reduce Exposures

Although steps that reduce airborne contaminant levels are usually preferred, here are some techniques to reduce exposures by changing the work patterns or protecting the employees:

• Administrative exposure controls that involve adjusting work schedules or rotating job assignments. In this way no employee works long enough in an area of high concentration to receive an overexposure.

- Personal protective equipment such as respiratory protective devices. There are a variety of respirators suitable for different contaminants and levels of exposures. They are divided into two main classes: air purifying respirators (Figure 1.7) that remove contaminants from ambient air; and atmosphere-supplying respirators that provide respirable air to the worker from an air compressor, large compressed air cylinder, or, in the case of a self-contained breathing unit (Figure 1.8), from an air cylinder carried by the worker. Generally, respirators are used for routine tasks only when engineering or administrative controls do not reduce exposures to acceptable levels. Respirators may be needed during emergencies or unusual conditions. Federal and state Occupational Safety and Health regulations contain specific requirements for respirator use.[6] Follow these standards when establishing a respirator program.[7]
- Training and education that help employees reduce exposures. If contaminants do not have a noticeable odor or other sensory warning properties, employees may not understand how exposures occur. If employees have this information as well as knowledge of the contaminants' effects, they may be able to help devise methods to reduce their exposures.

WHEN ARE CONTROLS NEEDED?

The single biggest factor in current efforts to reduce employee exposures has been OSHA—the Occupational Safety and Health Act. The allowable

Figure 1.7   Air purifying respirators, such as this half mask for dusts, remove contaminants from the ambient air.

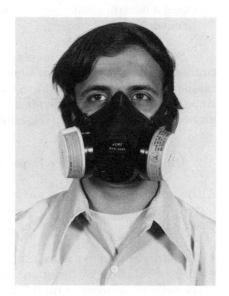

Figure 1.8 This self-contained breathing apparatus is an example of atmosphere-supplying respirators. Breathing air is provided from a safe source independent of the air around the worker.

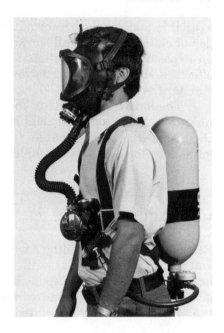

chemical exposure limit and other standards adopted by the U.S. Department of Labor require ventilation under specific conditions and set exposure limits for over 400 different chemicals. Similar standards were adopted by the states that have their own OSHA programs.

OSHA Standards

OSHA standards relating to ventilation are covered in Chapter 3. Except where ventilation is specifically mentioned, such as for welding on cadmium, lead, or other toxic metals, ventilation or another exposure control technique is needed if sampling shows that employee exposures exceed allowable levels, if an oxygen deficiency could occur, if build-up of flammable or explosive materials exceed safe limits, or if other evidence of harmful exposures exists. Most chemical exposure standards are based on an employee's average exposure over the work shift. Some standards set a maximum allowable "ceiling" level. A few standards even permit a short-term peak exposure above the allowable ceiling concentration.

Minimizing Exposures

OSHA standards do not exist for all chemicals and other substances used in the occupational environment. Current OSHA standards are set using available toxicological information along with a safety factor, but they do not claim to protect all workers in all exposure situations. Whenever an exposure causes significant discomfort or apparent adverse health effects, it should be reduced even if no OSHA numerical health exposure standard is exceeded.

The long-term effects of many compounds on the human body are unknown. It is good industrial hygiene practice to minimize exposure to chemicals whenever possible. Chapter 4 (Hazard Assessment) contains guidelines for evaluating employee exposures to airborne materials.

WHO IS RESPONSIBLE?

Responsibility for providing a work environment free of recognized safety and health hazards rests with management. Since ventilation systems help in this effort, the responsibility for deciding whether ventilation is needed, for designing and installing it, and for using and maintaining the system is usually delegated by management.

Safety or Industrial Hygiene Department

In many plants the safety or industrial hygiene department has responsibility for OSHA compliance and for identifying health and safety hazards. This staff is generally trained in determining worker exposures or is able to obtain these services from outside the plant. The staff may also be responsible for using velocity or pressure measurements to test the performance of ventilation systems. Enforcing the use of ventilation, respirators, or other exposure controls is another typical safety department function.

Plant Engineering

The plant engineering or technical services department usually has the responsibility for designing and installing exhaust ventilation systems. They have the understanding of the industrial processes needed to identify the origin of emissions, and they have the knowledge to design system hardware. They or the maintenance department may be responsible for the periodic inspection and maintenance needed to keep the system performing effectively.

Production Department

Line managers and supervisors should be responsible for making sure ventilation is used when needed and for scheduling jobs so that operations needing ventilation are performed in areas or on machines with adequate airflow. They also should be responsible for training the workers to use the ventilation system effectively.

Worker Involvement

Since the ventilation system will be used by the workers, they should be involved in system planning and layout. An important criterion for any system is that it be usable once installed.[8] Ventilation hoods that interfere with the work may not be used at all. If hoods and ducts are located so that they are exposed to damage from forklift trucks or overhead cranes, workers may be able to warn the designers before damage occurs.

DILUTION VENTILATION SYSTEMS

As mentioned previously there are two types of exhaust ventilation: dilution ventilation (also called general ventilation), which dilutes the contaminants in the workroom air to acceptable levels; and local exhaust ventilation, which captures or contains contaminants at their source before they disperse into the workroom. Some advantages and disadvantages of these systems were also discussed earlier in this chapter. Since the bulk of this book focuses on local exhaust ventilation, the remainder of this chapter will cover the application and design of dilution ventilation systems.

Dilution, Not Removal

It is easy to picture air moving through the work area in a straight path from the air inlet to exhaust fan, almost as if traveling inside an invisible duct (Figure 1.9), to whisk contaminants out of the workroom. However, this does not usually occur. The incoming air diffuses throughout the room. Some of it passes through the zone of contaminant release and dilutes the contaminants to a lower concentration. The dilution continues as the material moves farther from the process until the contaminated air is removed by the exhaust fan. Depending on the location of the air inlet and exhaust fan, and the total airflow through the room, a consid-

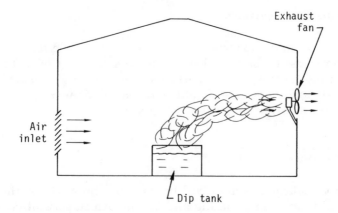

Figure 1.9 Incorrect visualization of dilution ventilation. The contaminants actually disperse throughout the right half of the room as shown in Figure 1.1.

erable time period may elapse after the process stops before all contaminants are removed from the room. Dilution occurs from natural
ventilation as well as from mechanical systems that use fans or other
air-moving devices.

Natural Ventilation

Natural ventilation is air movement within a work area due to wind,
temperature differences between the exterior and interior of a building,
or other factors where no mechanical air mover is used.

Even moderate winds can move large volumes of air through open
doors or windows. A 15 mile/hr wind blowing directly at a window with
an open area of 36 ft² can move about 25,000 ft³/min through the window
if the air can escape from the building through another opening. This
may be enough dilution airflow if the wind is reliable or if production
can be scheduled to coincide with favorable winds and the building is
not shielded from the wind by trees, hills, or other structures. The problem is that in many parts of the country this large dilution air volume
must be heated in winter, and fuel is expensive.

Air movement due to temperature differences may be more useful
than motion caused by wind. Hot processes heat the surrounding air and
the rising column of warm air will carry contaminants upward. Roof
ventilators allow escape of the warm air and contaminants (Figure 1.10).
As long as a worker does not have to lean over the heated process and
breathe the rising contaminated air, this type of natural ventilation may
be adequate. A good supply of replacement air for the building is needed,
especially during winter when doors and windows may be closed to minimize drafts.

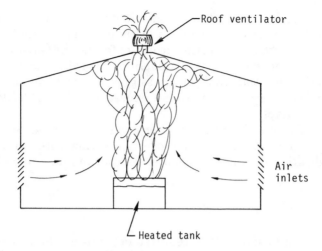

Figure 1.10 Natural ventilation as heated air rises around a hot process. This is not acceptable if a worker bends over the tank and breathes the contaminants.

Mechanical Ventilation

Wall-mounted propeller fans are often used for dilution ventilation (Figure 1.1) to provide a constant, reliable flow of air. These fans are described more fully in Chapter 9 but a major characteristic is that they are efficient air movers as long as a ready supply of replacement or make-up air can enter the area being exhausted. An adequate air supply system or sufficient air inlets are required for dilution ventilation systems.

Evaluating Dilution Ventilation

Deciding whether dilution ventilation is a good choice depends on these factors:

- The air volume needed to dilute the contaminants to safe levels may be excessive if large quantities of contaminants are released.
- Sufficient dilution must occur before workers inhale contaminated air. If employees work close to the contaminant source, the dilution airflow may have to be increased to reduce concentrations to safe levels before the air reaches the employees' breathing zones. This can be a real problem in manual gluing or surface-coating operations where workers bend over the work and breathe solvent vapors. With dilution ventilation for fire protection the dilution must occur before the contaminants reach a source of ignition.

- Only low fire hazard or low toxicity materials should be considered for control by dilution ventilation. Although there is no firm toxicity classification system, the American Conference of Governmental Industrial Hygienists uses the guidelines in Table 1.1 based on the Threshold Limit Values (TLV's) assigned to chemical substances as an indication of safe occupational exposure levels.[1] TLV's are explained in Chapters 3 and 4.
- The rate of contaminant release, or evolution, should be reasonably constant. This avoids the need for high airflow rates to provide adequate dilution during periods of peak contaminant release.

Calculating Dilution Airflow

The amount of dilution airflow required depends on the rate of contaminant released, its toxicity or flammability, the acceptable airborne concentration, and the relative efficiency of the total air volume flowing through the area in diluting the contaminants. The room size is not used to calculate dilution requirements since the airflow for these systems is not based on the "number of room air changes per hour" as is often used for general comfort exhaust ventilation.

Dilution Ventilation for Health

The equation for calculating the dilution airflow rate for toxic or irritating contaminants is: [1]

$$Q = \frac{F \times sp\ gr \times W \times K \times 1,000,000}{M \times L} \tag{1.1}$$

where Q = dilution airflow, ft³/min
    W = amount of liquid used per time interval (see Table 1.2)
    F = conversion factor for units of W, from Table 1.2
 sp gr = specific gravity of liquid (water = 1.)
    M = molecular weight of contaminant
    L = acceptable airborne concentration of contaminant, usually from OSHA standards or TLV list, ppm
    K = dimensionless safety factor to increase the calculated airflow rate over the minimum, in order to take nonideal conditions into account. K normally ranges from 3 to 10 depending on the number of workers exposed, overall effectiveness of the ventilation system, and uniformity of contaminant evolution. A higher K value is associated with poor airflow conditions and larger numbers of exposed workers.

Table 1.1   Toxicity Guidelines for Dilution Ventilation[1]

Toxicity Class	TLV Range, ppm
Slightly toxic	≥500
Moderately toxic	100–500
Highly toxic[a]	≤100

[a]Not recommended for dilution ventilation.

Table 1.2   Factors for Calculating Dilution Ventilation Airflow[a]

Amount of Liquid Used Per Time Interval (W)	Conversion Factor (F)
pints/hr	6.7
pints/min	403.
gallons/hr	53.7
gallons/min	3222.
liters/hr	14.1
liters/min	846.

[a]For use in Equations (1.1) and (1.2)

Example: Adhesive containing 60% toluene as a solvent is used at the rate of 3 pints/ hr in a large open-bay work area. The general layout of the workroom prevents much of the dilution air from passing directly through the zone of contaminant evolution at the workbench. Calculate the dilution airflow requirement.

Answer: The amount of solvent released is 60% of 3 pints/hr, or 1.8 pints/hr. The specific gravity of toluene is 0.87 and its molecular weight is 92.[10] The acceptable airborne concentration for toluene is 100 ppm from the TLV listing [11], although this should be checked against current OSHA standards. Assume a K factor of 8 since the air distribution through the zone of contaminant release is relatively poor. From Table 1.2 an F factor of 6.7 is used in Equation 1.1 when solvent evaporation is expressed in units of pints/ hr. Applying Equation 1.1:

$$Q = \frac{6.7 \times 0.87 \times 1.8 \text{ pints/hr} \times 8 \times 1{,}000{,}000}{92 \times 100}$$

$$Q = 9124 \text{ ft}^3/\text{min}$$

As this example illustrates, the magnitude of the safety factor K has a direct influence on the airflow calculation. Unfortunately K is ambiguous and it is difficult to assign it a meaningful value. Thus it is difficult to be confident that exposure tests after the dilution system is installed will show that exposures have been reduced sufficiently.

Dilution for Fire Protection

Dilution ventilation is used to reduce concentrations of flammable or explosive gases, vapors, or dust to safe levels well below their lower explosive limit (LEL). The dilution must occur before the contaminated air reaches any source of ignition. The accumulation of flammable or explosive mixtures in basements, pits, and other locations also must be considered in addition to diluting vapors in the general work area.

The equation for calculating dilution airflow for fire/explosion prevention is:

$$Q = \frac{F \times sp\ gr \times W \times C \times 100}{M \times LEL \times B} \qquad (1.2)$$

where Q = dilution airflow, ft^3/min
W = amount of flammable liquid used per time interval (see Table 1.2)
F = conversion factor for units of W, from Table 1.2
$sp\ gr$ = specific gravity of liquid (water = 1.)
C = dimensionless safety factor that depends on the percentage of LEL acceptable for safe conditions. For some applications the concentration should not exceed 25% of the LEL so C = 4; for other situations C values of 10 or higher may be needed.
M = molecular weight of contaminant
LEL = lower explosive limit of contaminant, percent
B = constant reflecting that the LEL decreases at elevated temperatures. B = 1 for temperatures up to 250°F, B = 0.7 for temperatures above 250°F.

Some operations release peak amounts of contaminants over a short time period. For example, drying ovens evaporate solvents rapidly during the first few minutes after objects are placed in the oven. The value selected for W should reflect peak release rates.

Equations 1.1 and 1.2 both assume that the dilution air has "standard" density of 0.075 lb/ft^3. This is the density of air at 70°F, 29.92 in. of mercury atmospheric pressure and 50% relative humidity. Factors that affect density are temperature, altitude, and humidity. Density correction calculations are explained in Chapter 9. For dilution systems the most common factor is high temperature within drying ovens or similar enclosures. Density adjustments for high temperatures can be calculated from:[1]

$$Q_{actual} = Q_{calc} \left(\frac{460°F + T}{530°F} \right) \qquad (1.3)$$

where Q_{actual} = dilution airflow at actual temperature, ft³/min

Q_{calc} = dilution airflow calculated from Equations 1.1 or 1.2, ft³/min

T = actual dilution air temperature, °F

Example: One gallon of toluene evaporates per hour in an adhesive drying operation at 200°F. Observations show that most of the solvent evaporates within the first 10 minutes of the drying cycle. How much airflow is needed to keep the concentration below 20 percent of the LEL?

Answer: Since one gallon evaporates within about 10 minutes, the value of W is 0.1 gal/min. From Table 1.2 the value of F is 3222. The specific gravity of toluene is 0.87, the molecular weight is 92 and the LEL is 1.3%.[10] Since the temperature is less than 250°F, B = 1. To maintain the concentration below 20% of the LEL, C = 5. Applying Equation 1.2:

$$Q = \frac{3222 \times 0.87 \times 0.1 \times 5 \times 100}{92 \times 1.3 \times 1}$$

Q = 1172 ft³/min

Since this operation is at 200°F, adjust the calculated airflow using Equation 1.3.

$$Q_{actual} = Q_{calc} \left(\frac{460°F + T}{530°F} \right)$$

$$= 1172 \text{ ft}^3/\text{min} \left(\frac{460 + 200}{530} \right)$$

$$= 1172 (1.25) = 1465 \text{ ft}^3/\text{min}$$

When both employee exposure and fire/explosion prevention are considered for the same operation, the dilution flow rate calculated using Equation 1.1 usually governs because the allowable airborne levels for breathing are significantly lower than the LEL's for almost all, if not all, substances.

Dilution Ventilation System Layout

Dilution systems work best when the air inlet and exhaust fan are located so that as much air as possible flows through the zone of contaminant release. Only the air that passes through the area where contaminants are released is available for immediate dilution of contaminants to safe levels. Air supply systems with blowers and air outlets near the work area help to provide the correct amount of air at the right place.

The plant should be arranged so that air movement is from cleaner to dirtier areas. Locate the processes or locate the fan so that the units that release contaminants are as close as possible to the fan. Also eliminate or provide separate exhaust for areas where contaminants can accumulate and defeat the dilution effect of the airflow from the overall system.

SUMMARY

Ventilation is an effective way to control toxic or flammable contaminants. There are two types of ventilation for contaminant control: dilution ventilation that reduces contaminant concentrations by diluting them with fresh air, and local exhaust ventilation systems that capture or contain contaminants at their source before they are dispersed in the workroom.

Although both types of ventilation are useful, local exhaust is usually preferred, if it is feasible. Advantages include more positive control of employee exposures and lower overall airflow requirements. On the other hand, dilution ventilation systems are usually less expensive to install and operate if heating costs during the winter are not excessive.

This book focuses on how local exhaust systems work and how to use them in your plant for controlling contaminants. It is intended for safety professionals, plant engineers, industrial hygienists, and others with a role in using ventilation to protect health.

REFERENCES

1. ACGIH Committee on Industrial Ventilation. *Industrial Ventilation—A Manual of Recommended Practice,* 17 Ed. (Lansing, Michigan: American Conference of Governmental Industrial Hygienists, 1982).
2. American Industrial Hygiene Association. *Heating and Cooling for Man in Industry* (Akron, Ohio: AIHA, 1975).
3. *Federal Register* 38, No. 66, 8820 (1973).
4. Olishifski, J.B. and F.E. McElroy, Eds. *Fundamentals of Industrial Hygiene* (Chicago, Illinois: National Safety Council, 1971).
5. Bastress, E.K., J.M. Niedzwecki, and A.E. Nugent. "Ventilation Requirements for Grinding, Buffing and Polishing Operations," U.S. Department of Health, Education and Welfare, Publication No. (NIOSH) 75–107 (Washington D.C.: U.S. Government Printing Office, 1975).
6. OSHA General Industry Safety and Health Regulations, U.S. Code of Federal Regulations, Title 29, Chapter XVIII, Part 1910.134 (1975).
7. Ruch, W.E. and B.J. Held. *Respiratory Protection—OSHA and the Small Businessman* (Ann Arbor, Michigan: Ann Arbor Science Publishers, Inc., 1975).

8. National Safety Council. "Checking Performance of Local Exhaust Systems," Data Sheet 438 (Chicago, Illinois: National Safety Council, 1963).

9. Hemeon, W.C.L. *Plant and Process Ventilation* (New York, New York: Industrial Press, Inc., 1963).

10. Sax, N.I. *Dangerous Properties of Industrial Materials* (New York, New York: Van Nostrand Reinhold Company, 1975).

11. American Conference of Governmental Industrial Hygienists. "Threshold Limit Values for Chemical Substances in Workroom Air" (Cincinnati, Ohio: ACGIH, 1983).

CHAPTER 2

Indoor Air Pollution: Ventilation as Source and Solution

The quality of air in residential, commercial, and office buildings is an important issue for building managers, engineers, and health professionals. Ventilation can play a vital role in maintaining air quality by controlling contaminants. Dilution of contaminants by increasing ventilation is usually the most practical approach. At the same time, the current trend towards reduced ventilation for conserving energy has contributed to air quality complaints.

INDOOR POLLUTION PROBLEMS

HVAC System Objectives

Heating, ventilating, and air conditioning (HVAC) systems for buildings are designed to meet three objectives:[1]

- Good air quality—adequate oxygen levels, removal of carbon dioxide resulting from human respiration, and dilution/removal of contaminants including odors.
- Thermal comfort—typically in an acceptable thermal environment at least 80 percent of normally clothed occupants engaged in sedentary or near-sedentary activites feel comfortable.[2] Environmental

factors contributing to thermal comfort include air temperature, radiant heat load, relative humidity, and air velocity. In modern sealed buildings, maintaining sufficient air motion with low overall ventilation rates can be a challenge.

• Reasonable energy costs—approximately 8% of the total energy consumed in the United States is used to heat, cool, and ventilate institutional and commercial buildings.[3] Since heating or cooling outdoor air as it enters a building is a major cost, designers have minimized the quantity of outdoor air used for ventilation, especially during periods when the building is unoccupied. Steps to seal buildings to reduce infiltration also minimize the quantity of outdoor air entering modern buildings.

Tight Building Syndrome

These three objectives provide clues to how building problems can develop. Soaring energy costs lead to reexamination of established ventilation airflow standards and the ways the HVAC systems are operated. Reducing the amount of outdoor air can prevent adequate dilution of indoor-generated contaminants. Thermal comfort can also suffer, since lower air circulation rates make it difficult to maintain adequate air motion. These changes in HVAC design to meet the national goal of energy conservation occurred without a full understanding of indoor-generated contaminants, including emissions from building materials and furnishings in new buildings. What can follow are complaints of poor air quality and symptoms of eye and respiratory tract irritation, headache, or allergic responses. The effects of these episodes of acute indoor pollution are characterized as tight building syndrome.

These episodes are often difficult to diagnose because of confounding factors:

• Airborne levels of specific contaminants are typically well below published allowable levels for the occupational setting. No single contaminant can be identified as the cause of the complaints.

• Individual health problems, stress from the use of video display terminals, management-employee relations, and other factors may be involved.[4]

• Widespread episodes may be triggered by an accumulation of smaller problems that remain unsolved. Thus a cause-effect relationship, valuable in diagnosing problems, may be absent, making it difficult to identify the extent and cause of the complaints.

Chronic Indoor Exposures

Since people spend up to 80–90% of their time indoors,[2] the overall issue of indoor air quality is much broader than the smaller subject of solving acute problems. Recognition of the contribution from indoor exposures to the total dose of pollutants that people receive has encouraged air pollution and public health researchers to expand outdoor pollution measurements and health criteria to include indoor levels. They now use indoor/outdoor (I/O) pollution ratios to relate indoor levels of contaminants to the ambient (outdoor) levels. Studies of I/O ratios help to identify predominantly indoor sources (I/O > 1) and assist in predicting the effect of long-term indoor exposures to the general population. In studies of specific buildings these ratios must be used with care, since the I/O ratio will change if only the outdoor level varies.

Chronic exposures, by definition, do not cause immediate sensory response or acute symptoms and are not considered part of tight building syndrome problems.

TYPICAL CONTAMINANTS

Published studies on indoor air pollution have identified at least eight major contaminants: five of them (carbon monoxide, formaldehyde, organic vapors, respirable particulate matter, and microorganisms) can play a role in acute outbreaks while the remaining three (asbestos, radon and its progeny, and carbon dioxide) plus formaldehyde have long-term or chronic health implications. These substances are listed in Table 2.1 along with a summary of typical indoor levels, air monitoring techniques, and other comments. Refer to Chapter 3 (OSHA Ventilation Standards) and Chapter 4 (Hazard Assessment) for an explanation of the terms used in Table 2.1.

Carbon Monoxide

Carbon monoxide (CO) from faulty furnaces or automobile exhaust is a well-known hazard. Simple detector tubes or a direct-reading monitoring instrument should be used to evaluate any situation where CO could be present and acute symptoms resemble CO poisoning (Table 2.1).

In buildings where heavy smoking occurs with poor ventilation, indoor CO levels may rise to about 25–35 ppm. Current ventilation standards recognize that significantly more airflow is needed in smoking areas than in non-smoking areas. However, this need is based primarily on achieving acceptable odor levels rather than controlling CO levels.

Table 2.1 Typical Indoor Contaminants in Office and Commercial Buildings

Contaminant	Indoor Source	Levels	Implicated in Acute Symptoms?	Air Monitoring Techniques	Comments
Carbon Monoxide	Faulty furnaces, cigarette smoking, attached garages	See Comments	Yes—faulty furnaces or attached garages can cause headache, nausea, etc. No— sidestream cigarette smoke	Detector tubes Electrochemical or other electronic instruments	CO levels up to 25 ppm have been found in heavy smoking areas. Above this level other causes should be suspected.
Formaldehyde	Building materials (insulation, particle board, etc.), furnishing and fabrics	0.05–1.0 ppm	Yes—symptoms are eye and respiratory irritation, headache, nausea, fatigue and thirst	Passive dosimeters Bubbler with sampling pump	Animal studies have implicated formaldehyde as a carcinogen.
Organic Vapors (besides Formaldehyde)	Solvents, resins, aerosol sprays		Yes—complaints of odors, symptoms are eye and respiratory irritation, and headache	Activated carbon or other adsorbent with sampling pump Passive dosimeters Direct reading instruments Detector tubes	Levels may be too low to detect with routine techniques. Special consultants can provide detailed analysis to parts per billion level if needed.
Respirable Particulate Matter	Smoking, aerosol sprays, dust resuspension	100–300 $\mu g/m^3$	Yes—symptoms are eye and respiratory irritation	Respirable dust cyclone or impactor with pump	None

Contaminant	Source	Typical Levels	Related to Ventilation	Measurement Method	Comments
Airborne Microorganisms	Infected occupants, dust, contaminated HVAC systems, molds and fungi	Not applicable	Yes—allergic response and infection	Direct reading instruments (laser counter, etc.) Air sampling with laboratory culturing to identify agents	Often difficult to identify organism causing problem.
Asbestos	Insulation and other building materials containing asbestos	See *Comments*	No	Analysis of bulk material to identify asbestos Air sampling with filter and pump followed by counting of fibers using phase contrast microscopy	Airborne levels are often episodic following disturbance of asbestos-containing materials. Routine sampling may not detect episodic exposures.
Radon and its Progeny	Building materials, soil and ground water	0.1–30. nanocuries per m^3	No	Passive dosimeter containing special plastic element that registers alpha particles for counting in the laboratory	None
Carbon Dioxide	Metabolic activity of occupants	Up to 3000 ppm	No	Infrared or other direct reading instruments Detector tubes	About 400 ppm is typical for ambient air. Elevated indoor levels may indicate too many occupants for the HVAC system.

Studies of CO levels in buildings near freeways show that high CO concentrations can occur if HVAC systems with fresh air intakes near ground level are set to draw outdoor air into the building during the morning or evening commute period.[5]

Formaldehyde

Formaldehyde may cause acute symptoms when it is present at levels found in some problem buildings. Since studies link formaldehyde to cancer in laboratory animals, it may be a long-term problem as well. Formaldehyde is a component in urea-formaldehyde (UF) resin, which is used in building materials such as UF foam insulation. It is also used as a plywood adhesive, a binder in particle board, and a component in floor coverings, fabrics, and other furnishings used in commercial or industrial buildings. Other sources of formaldehyde include unvented combustion (gas stoves) and cigarette smoke.[5]

New construction or newly refurbished buildings tend to have higher formaldehyde levels than do older buildings. The emission rate decreases with the age of the source, but one study showed some emissions after 16 months following installation of UF foam insulation in wall panels.[6]

Organic Vapors

Organic vapors are discussed as a general category because, except for formaldehyde, individual organic compounds tend to be found at levels far below concentrations that would be expected to cause health problems. There is also a large number of separate compounds present; identifying and quantifying each one is extremely difficult, if not impossible. Some of these compounds may be lacrimating (tearing) or particularly irritating agents, even at extremely low levels. Major sources of organic contaminants are:

- Bioeffluents—humans emit organic substances from natural biologic processes. Compounds such as methanol, ethanol, acetone, and butyric acid were identified in a school auditorium seating 400 people.[2] Typically the HVAC system would keep bioeffluents at acceptable levels.
- Building materials—adhesives, paints, sealants, carpeting, and other materials release organic vapors, especially when new. Since emission rates decline as the building "cures," steps such as higher temperature settings and increased airflow have been tried to speed curing while maintaining acceptable air quality during the process.

- Consumer products–spray cleaners, pesticides, photocopy machine solvents, and other materials all add to the overall level of organic vapors.
- Tobacco smoke–contains a variety of organic vapors in addition to the CO discussed earlier.

Sometimes measurements of total hydrocarbons or non-methane hydrocarbons (NMHC) are performed indoors. Both of these analyses are widely used in ambient air pollution but must be evaluated with caution indoors since indoor values may be compared to the National Ambient Air Quality Standard (outdoor pollution standard) for NMHC. It is 160 $\mu g/m^3$ averaged over the 3 hours between 6:00 A.M. and 9:00 A.M. but is based on NMHC as a precursor of smog and is not directly related to health effects. In some indoor studies this 160 $\mu g/m^3$ level has been exceeded.[3]

Respirable Particulate Matter

Respirable particulate matter (RPM) is another broad category of indoor pollutants that includes particles, such as dusts and pollen, and aerosols in the size range that can reach the lungs when inhaled, as explained in Chapter 6. Although these pollutants vary in chemical composition, they are grouped together because air sampling devices collect particulates according to size distribution. High RPM levels can cause eye and respiratory tract irritation.

For many people the main exposure to RPM is from tobacco smoke.[2] Smokers receive a higher total dose than nonsmokers but sidestream smoke is produced as long as the tobacco burns, and many nonsmokers are more sensitive to these substances than are smokers. Smoking can cause an increase of a factor of 3–40 in RPM over background levels.[7] Other sources of RPM are aerosol sprays and resuspension of dust.

Practically all HVAC systems have a filter in the system to control particulates. These filters may not be adequate when RPM levels are very high. If the source of RPM is confined to a relatively small area, local exhaust systems or local air cleaners such as portable electronic air cleaners may be helpful. Limitations on smoking may be a solution when tobacco smoke is a major source of RPM.

Airborne Microorganisms

Inhalation of discharged aerosols from infected people is a main mechanism for transmitting acute respiratory infections. Where these diseases are transmitted by person-to-person contact at close range, it is not clear whether reducing ventilation rates increases the incidence of disease. If the infectious agents remain viable and airborne for extended periods, adequate ventilation rates and filtration of recirculated air should reduce the incidence. Tuberculosis, measles, and staphylococcus infections are known to be transmitted by ventilation systems in schools and hospitals[8]; therefore, adequate treatment (filtration or disinfection) can be an important control.

Some studies in problem buildings implicate the HVAC system itself as a source of microorganisms.[9] Filters that have not been cleaned and microbial growth on damp surfaces in the air conditioner portion of the system cause airborne microorganisms to be distributed throughout the building. The first steps in evaluating problem buildings are a thorough inspection of the HVAC system and a thorough cleanup.

Asbestos

Asbestos in building materials has received wide publicity. Federal, state, and local laws govern its use and control removal of existing asbestos-containing material.

Dilution ventilation is generally not acknowledged to be a primary solution to airborne asbestos in buildings.

Radon and Its Progeny

Building materials, soil around buildings, and water may release radon gas through radioactive decay of naturally occurring radium-226. The radon in turn decays to form short-lived radioactive decay products (called radon daughters or progeny) that include polonium, lead, and bismuth. These materials deposit in the lungs of building occupants either directly or by attaching to airborne particles, which are then inhaled.

During the radioactive decay process, alpha particles, which pose a risk of increased lung cancer at some (unknown) level, are emitted. Of course, this process has been occurring since the beginning of time, but current concern is whether weatherization and reduced airflow rates allow a buildup of radon and its progeny in the air to a level that significantly increases the risk to the population. Uranium miners with working lifetime exposures to levels thought to be 100–100,000 times higher than typical indoor concentrations do have increased incidence of lung cancer. However, the extent of indoor measurements is extremely limited, and radon and its progeny will be a major issue in the future if indoor levels are found to be a significant source of exposure to radiation.

Since there are no acute symptoms, radon and its progeny are not factors in tight building syndrome outbreaks. To dilute the levels that occur from natural decay, however, ventilation is a potential solution.

Carbon Dioxide

Carbon dioxide (CO_2) is present in the ambient environment at levels averaging about 400 ppm. The main indoor source of CO_2 is the metabolic activity of occupants; smoking and wood-burning fireplaces do not increase levels significantly.[5]

As buildings are weatherized and ventilation rates are reduced, some ventilation engineers feel that CO_2 buildup may be an important factor in setting adequate ventilation standards for energy efficient buildings.[5] A technique for estimating building ventilation rates by using measurements of occupant-generated CO_2 as a "tracer gas" is described later in this chapter.

Humidity

Water vapor is not listed as a contaminant in Table 2.1 but it affects indoor air quality in several ways:

- Low relative humidity can cause particles and vapors in the air to be more irritating than at high levels.
- High humidity aggravates odor problems and favors mold growth. High humidity leads to complaints of stuffiness and thermal discomfort.

Excessive relative humidity in air-conditioned buildings is controlled by chilling the supply air to below its dew point to condense the moisture, then reheating the air for thermal comfort.

CONTROL TECHNIQUES

Although this chapter focuses on HVAC systems and indoor air pollution problems, ventilation is not the sole solution, nor in many cases the most effective. For some contaminants, such as asbestos, ventilation is not even recommended in most situations. Other controls are:

- Source removal or substitution.
- Source modification, which usually means the use of sealants or barriers to prevent emissions to the atmosphere.
- Air cleaning devices, either installed in the HVAC system to clean recirculated air or located near the contaminant source to clean the nearby air. Air cleaners are covered in more detail at the end of this chapter and in Chapter 7.
- Local exhaust ventilation systems to control a specific machine or other discreet source of air contaminants.

- Administrative controls, such as limitations on smoking or use of solvent-based spray cleaners, that the employer or building manager can enforce through procedures such as work rules or contracts. When these measures cannot be enforced and so are voluntary, this approach is called *behavior modification*. Regardless of the name, success depends on educating the building occupants to the need for changing established patterns in order to improve air quality and conserve energy.

The application of specific control techniques for the contaminants discussed earlier is summarized in Table 2.2. It is important to realize that Table 2.2 is general and may not apply to specific situations. However, this table helps to illustrate that ventilation is often not the best answer.

EVALUATING THE PROBLEM

Acute indoor air problems can surface through complaints about odors, headache, nose and throat irritation, or similar symptoms. Often the symptoms begin with a few individuals and are easy to overlook in the early stages. Where the existence of a problem is not recognized and dealt with promptly, employees' health concerns and frustration at being ignored can give the situation an emotional element that makes a purely technical solution difficult to achieve. Evaluating the problem, once it is identified, usually begins with questionnaires or interviews with employees to determine the type and extent of symptoms and their commonality throughout the building. It is important to identify when (time, day of week, or season) symptoms change and what other factors employees feel may contribute to the problem. If it appears that reported symptoms may be explained by hay fever, the flu season, or problems unrelated to air quality within the building, it may be helpful to repeat the questionnaires or interviews in a control building with no reported indoor air quality problem. In this way it may be possible to determine whether the incidence of symptoms within each building is comparable.

Once the type, extent, and locations of symptoms have been established, finding the cause is a step-by-step procedure that involves:

- Sufficient knowledge about the HVAC system's components, layout, and function to determine whether the system is a possible cause of the problems. This may require a review of controls, damper operation, and cleanliness of system internals, which is discussed later in this chapter.
- Air sampling to measure levels of contaminants that may be involved. This is explained as part of the hazard assessment process in Chapter 4.

Table 2.2 Summary of Control Strategies for Indoor Contaminants

Control Technique	Carbon Monoxide	Formaldehyde	Organic Vapors	RPM	Microbes and Allergens	Asbestos	Radon and Progeny	Carbon Dioxide
HVAC System	A	A	A	A	A	N	A	A
Source Removal/ Substitution	A	A	A	A	A	A	A	N
Source Modification (sealing, etc.)	A	A	A	A	A	A	A	N
Air Cleaning Devices	N	A	A	A	A	N	N	N
Local Exhaust	A	N	A	A	N	A	N	N
Administrative Controls/Behavior Modification	A	A	A	A	A	A	N	N

A: Generally applicable or recommended.
N: Generally not applicable or recommended.
Source: References 2 and 8.

- Ventilation airflow and, if appropriate, infiltration rate measurements to determine whether air exchange rates, air motion, temperature, and relative humidity are within acceptable criteria.

What Are Acceptable Ventilation Rates?

Ventilation criteria were established soon after mechanical ventilation systems were developed. The early strategy assumed that maintaining a relatively odor-free environment would also take care of unknown but possibly harmful airborne substances. Pioneering studies by Yaglou[10] in the 1930s measuring odor acceptability in a controlled experimental chamber led to the concept that acceptable criteria were based not only on airflow (ft³/min) per person but also on the occupancy load in the building. For example, 7.5 ft³/min per person was acceptable with 3 people in the test chamber, but 25 ft³/min per person was needed for an acceptable environment with 14 people in the chamber. The required ft³/min per person value increased with occupancy load in order to maintain an acceptable odor level.

Over the years the American Society of Heating, Refrigerating, and Air Conditioning Engineers (ASHRAE) has been the leader in translating experimental research and practical experience into voluntary or consensus standards.[11, 12] Their standards, in turn, are the basis for many state and local codes and so become more than voluntary standards. Since the ASHRAE standards are updated periodically, it is important to refer to the latest edition. ASHRAE standard 62-1981, "Ventilation for Acceptable Indoor Air Quality," is the most recent attempt to maintain air quality while conserving energy. The following major points are included in the standard:

- Separate ventilation rates for smoking and non-smoking occupancies are specified. The actual rates, which vary by type of occupancy (Table 2.3), show that up to five times as much air is needed when smoking occurs than when smoking does not occur. A minimum rate of 5 ft³/min is required for CO_2 removal.
- Quality of outdoor air is recognized as an important factor in maintaining adequate air quality. Numerical contaminant levels and other criteria for evaluating outdoor air quality are listed.
- Operation of the HVAC system to maintain adequate air quality is covered. For example, where contaminants evolve from building materials or other non-occupancy related factors, the lead time that the system must operate to remove contaminants before the daily period of occupancy begins is specified.

Since each ASHRAE standard is revised as more experience is gained, it is crucial not to make decisions based on an outdated version.

The airflow figures in Table 2.3 are listed only as an illustration; the current version of the standard should be consulted.

While current standards should be the basis of new HVAC system design, they are not always applicable in judging whether or not an existing system is adequate. Because of the high level of interest in indoor air quality, research findings are being published that are helpful in evaluating problems. For example, a recent study[7] found that with no smoking a ventilation rate of about 5 to 10 ft^3/min per person provided satisfactory odor levels for 75% of the test subjects entering an occupied chamber; however, a ventilation rate of 16 ft^3/min per person was needed to satisfy 80% of the subjects. Ventilation rates would have to be even higher to satisfy a larger percentage of the test subjects. This study also confirmed that high temperature (> 78° F) and high humidity (> 70° RH) worsened the odor problem. Results like these help explain why even relatively high airflow rates will not satisfy all occupants.

Measuring Building Ventilation Rates

Ventilation rates can be estimated by measuring airflow directly in the system ducts serving each part of the building, or by use of a tracer technique in which the gradual decline in the concentration of a tracer gas in the building is measured and related to air exchange rates.

Duct Measurements

Duct measurements are often used to evaluate the performance of local exhaust systems; the techniques are explained in Chapter 12. In com-

Table 2.3 Ventilation Rates for Smoking and Non-smoking Occupancies

	Outdoor Air Requirements, ft^3/min per person	
Occupancy	*Smoking*	*Non-smoking*
Retail Stores		
Showrooms	25	5
Warehouses	10	5
Theaters		
Lobbies	35	7
Offices		
Office space	20	5
Meeting space	35	5
Food and Beverage Service		
Dining Rooms	35	7
Bars	50	10

Note: These are listed as an illustration of the increased ventilation rates for smoking occupancies. Refer to the latest edition of the reference for current values.
Source: Reference 11.

mercial or office buildings, however, HVAC ducts may be difficult to reach for the required testing. In buildings with high ceilings, air discharged near the ceiling may not reach the occupied portion of the room; therefore, duct airflow values will not correlate with effective ventilation rates that benefit occupants. If outside air infiltration occurs through doors, windows, or other openings, duct readings will understate the actual airchange rate.

Measuring the airflow from all outlets supplying a work area is more convenient than duct measurements (Figure 2.1). However, some outlets (technically called *terminal devices*) use a jet of supply air or a small fan to draw in room air to induce extra air circulation in the room (Figure 2.2). Outlet readings on these units, unless adjusted for the quantity of room air being recirculated, will overestimate air supply rates. This is because a portion of the air being measured is recirculated room air rather than system air.

Tracer Techniques

Because of these limitations to direct duct measurements, tracer gas studies are widely used in indoor air quality evaluations. The tracer technique involves releasing into the room or building a stable, nontoxic, easily measured gas with a density near that of air; mixing the gas well; and then measuring its concentration periodically. The airborne level decreases with time due to dilution with supply air or infiltration of outdoor air through open doors or cracks. The concentration values, when plotted and analyzed as described in this section, provide an estimate of the air exchange rate.

Figure 2.1 Airflow measuring instrument for HVAC supply terminal units. *(Courtesy Shortridge Instruments, Inc.)*

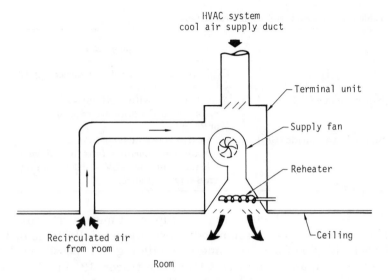

a. Recirculating terminal unit with fan

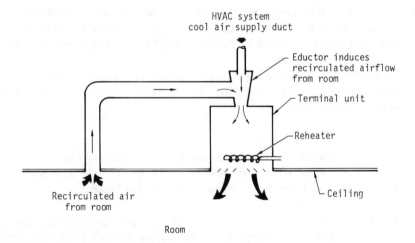

b. Recirculating terminal unit with induced flow

Figure 2.2 Terminal units that recirculate room air to increase air movement in the room. Airflow measurements at these terminal devices will overstate HVAC supply rates if not adjusted for recirculated air.

Although several gases have been used as tracers, none meets all the criteria for an ideal choice. Table 2.4 lists the characteristics of three gases that have been used: sulfur hexafluoride, nitrous oxide, and ethane. Because of the disadvantages of the tracer gases, it may not be possible

Table 2.4 Characteristics of Tracer Gases

Gas	Characteristics
Ethane (C_2H_6)	Explosive in concentrations exceeding 3% in air.
Nitrous oxide (N_2O)	Anesthetic gas with a NIOSH-recommended workplace exposure level of 25 ppm.
Sulfur hexafluoride (SF_6)	About five times heavier than air; usually need portable fans to achieve adequate mixing. TLV = 1000 ppm for an 8-hour exposure.

to conduct tests while the building is occupied. It may be difficult, if not impossible, to obtain permission to release a potentially toxic or flammable gas inside an office or commercial building. One possible solution is the use of occupant-generated CO_2 as the tracer.[13] In order to use CO_2, the indoor levels must be higher than ambient levels so that the dilution with outside air can be measured over time. While this is rarely the case in older buildings, it may occur in new tightly sealed buildings. These tracer tests are usually conducted after workers leave for the day to avoid interference caused by additional CO_2 that is contributed by the occupants during the test.

To perform a tracer study obtain the gas, a properly calibrated direct reading analytical instrument, and mixing fans, if needed. Then follow this general procedure:

- Make sure that the HVAC system controls are set to operate in a manner typical of the time period being evaluated. For example, in studies conducted after the close of business be sure that automatic setbacks will not reduce airflow rates if the tests are to represent normal daytime operations.
- Measure background levels of the tracer gas both indoors and outside.
- Release the gas and, when it is thoroughly mixed, make the initial reading and record it along with the time. Temperature, humidity, and system operating parameters should also be recorded so that the factors affecting ventilation rates can be reconstructed later.
- Periodically repeat the concentration measurements and record the level and time until the indoor level of the tracer drops to near ambient levels.
- Plot the concentrations as a function of time on semi-logarithmic paper (See Figure 2.3 used in the example problem). Draw a visual best-fit straight line through the data and read the initial and final concentrations and the elapsed time between the two. (A best-fit line is used to reduce the influence of atypical readings on the calculations).
- Calculate the air exchange rate using this equation:

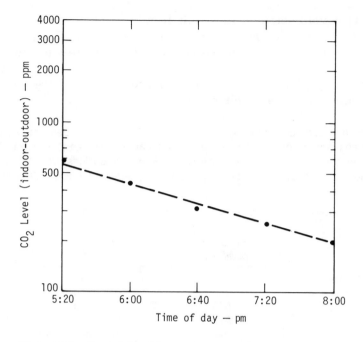

Figure 2.3 Semi-logarithmic plot of tracer gas concentration as a function of time.

$$A = \frac{1}{\Delta t} (\ln C_o - \ln C_t) \qquad (2.1)$$

where A = air exchange rate, air changes/hour
   C_o = indoor concentration of tracer gas at start of test
      (minus any level in outdoor air), ppm
   C_t = indoor concentration at end of test (minus any level
      in outdoor air), ppm
   Δt = elapsed time of test, hours

and ln is the natural logarithm (to base e).

• Convert the air exchange rate (A) into volumetric airflow by multiplying A by the room volume or, where the ceilings are high and complete mixing of incoming HVAC system air does not occur throughout the whole room, by the occupied volume in the lower part of the room as estimated during the study:

$$Q = A \frac{\text{air changes}}{\text{hour}} \times \text{Vol} \frac{\text{ft}^3}{\text{air change}} \times \frac{\text{hour}}{60 \text{ min}} \qquad (2.2)$$

where Q = volumetric airflow rate, ft³/min
   A = air exchange rate from Equation 2.1, air changes/hour
   Vol = room volume, ft³/air change (to allow units to cancel).

This value of Q includes ventilation and sources of infiltration that add outside air to the space being tested.

Example: A tracer study using occupant-generated CO_2 is to be performed in one large room in an office building.[13] The test will begin at quitting time (5:15 P.M.). CO_2 levels will be measured with an infrared instrument. Room dimensions are 100 ft × 50 ft × 10 ft; room occupancy is normally 45 people doing sedentary office work. Ambient CO_2 level is 325 ppm.

Answer: The following CO_2 levels are measured during the test:

Time	Indoor CO_2 Level, ppm	Inside Level-Outside Level, ppm
5:20 P.M.	905	580
6:00	775	450
6:40	655	330
7:20	575	250
8:00	525	200

The values in the first and third columns are plotted in Figure 2.3 and a best-fit line drawn. From this plot, the initial and final values as well as the elapsed time are read:

$$C_o = 550 \text{ ppm}$$
$$C_t = 200 \text{ ppm}$$
$$\Delta t = 2 \text{ hr } 40 \text{ min} = 2.67 \text{ hours}$$

Using Equation 2.1:

$$A = \frac{1}{\Delta t}(\ln C_o - \ln C_t)$$

$$= \frac{1}{2.67}(\ln 550 - \ln 200)$$

$$= \frac{1}{2.67}(6.31 - 5.30) = \frac{1.01}{2.67}$$

$$A = 0.38 \text{ air changes/hour}$$

The room volume is 50,000 ft³ (100′ × 50′ × 10′) so the volumetric airflow can be calculated from Equation 2.2:

$$Q = A \frac{\text{air changes}}{\text{hour}} \times \text{Vol} \frac{\text{ft}^3}{\text{air change}} \times \frac{\text{hour}}{60 \text{ min}}$$

$$= 0.38 \frac{\text{air change}}{\text{hour}} \times 50,000 \frac{\text{ft}^3}{\text{air change}} \times \frac{\text{hour}}{60 \text{ min}}$$

$$Q = 317 \text{ ft}^3/\text{min}$$

On a per-person basis this ventilation rate is:

$$\frac{317 \text{ ft}^3/\text{min}}{45 \text{ people}} = 7 \text{ ft}^3/\text{min per person}$$

This rate would satisfy ASHRAE Standard 62–1981 for a non-smoking area but not for a smoking area (Table 2.3).

Air Motion Is Important for Comfort

While the quantity of air supplied to a work area is vital for diluting odors and other contaminants, indoor air quality complaints may be related to comfort rather than contaminant levels.

Thermal comfort is a function of air motion, temperature, and humidity. When air lacks noticeable motion, it may feel up to 2 degrees Fahrenheit warmer than its actual temperature.[14]

Air motion is achieved through location and design of the terminal devices. They vary in design and complexity according to the system but usually fall into one of these categories: ceiling slot diffusers, perforated or louvered ceiling diffusers, circular ceiling diffusers, wall grills near the ceiling, and perimeter floor grills. Some outlets are designed to direct air along the adjacent ceiling or wall surface to avoid drafts, and to maximize the distance air is thrown from the outlet and the volume of room air entrained in the moving air to promote mixing.

When evaluating problem situations in which air motion might be the cause or a contributing factor, the following points should be considered:[2]

- Airflow patterns are different for heating and cooling cycles. Ceiling units are less desirable during heating since the buoyant warmer air stratifies along the ceiling. Likewise, floor outlets are less effective during cooling cycles.
- Good terminal device performance assumes that there are no obstructions to good room air circulation. Features that add flexibility to the office architecture, such as moveable walls or portable office partitions that extend to the floor, can impede air movement.
- In a modern, sealed building the beneficial effect of good air movement is limited if the temperature setpoint is higher than the range of comfort.

This means that if the temperature is too high, good air circulation will not necessarily make occupants feel comfortable. Instead, good air circulation eliminates the need to adjust temperatures lower than the value needed to achieve a high degree of occupant comfort. For example, almost no diffuser operating at a low volumetric flow rate to conserve energy can maintain enough air motion to minimize discomfort at 78°F for people in business suits.[4]

HVAC SYSTEMS

The design and operation of HVAC systems reflect their major role in overall energy consumption described at the beginning of this chapter. Various energy saving techniques are used to meet established airflow criteria for the minimum cost. HVAC systems for sealed buildings can be extremely complex depending on the cost of energy and climatic conditions. This section provides an overview so that proper evaluation of HVAC system mechanical operation and air distribution can be performed to identify indoor pollution problems.

Zoned Systems

In most buildings the space is divided into separate zones (Figure 2.4) that are either served by separate HVAC systems or, when part of a single HVAC system, have different terminal devices, controls, and ducting arrangements. Interior zones may not need the same heating and cooling capacity as exterior, or perimeter, zones since the perimeter zones are subject to solar heating in summer and loss of heat to the outdoors during winter. Since thermal gain or loss is not a major factor, some system designs or operating controls may limit the volumetric air flow to interior zones. Air quality investigations need to focus on whether problems are associated with a single zone or the whole building.

Single vs. Dual Duct Systems

Dual duct systems have one duct carrying cool air and a second duct carrying warm air (Figure 2.5). Both ducts feed into the terminal device,

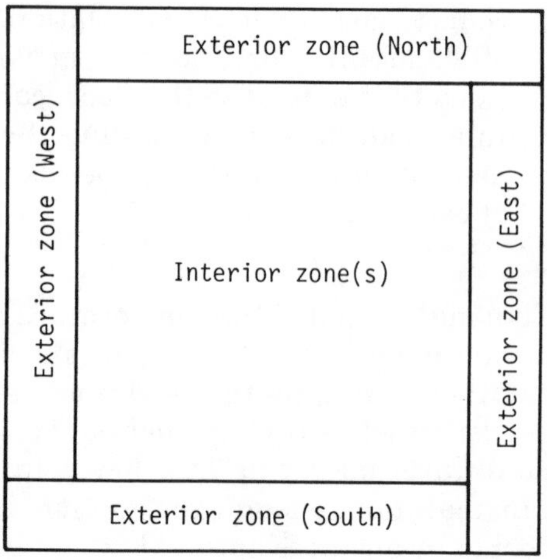

Figure 2.4   Typical zoned HVAC system. Since each zone has separate ducts and controls, it is important to identify which zone(s) are involved when evaluating problems.

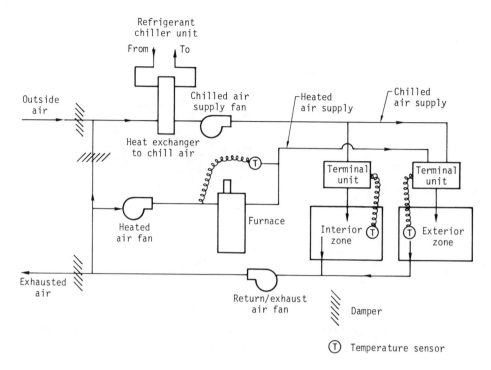

Figure 2.5 Dual duct HVAC system with separate ducts for heated and chilled air.

which proportions the amount of air from each duct depending on the thermostat setting in the room. Interior zones may only be served by the cool air duct since heat loss is low in these zones and heating capability is not needed. Needed warming in interior zones is provided by occupant activities or by air infiltrating from exterior zones.

A single duct system has only one duct that carries cool air (Figure 2.6). Heat is supplied to perimeter zones by local reheaters in the terminal units (Figure 2.2) or by baseboard or other heating devices in the rooms.

Constant Flow vs. VAV Systems

A constant flow system supplies a fixed volumetric airflow during periods when the building is occupied. Temperature is controlled either by proportioning the amount of cool or warm air (dual duct) or by reheating the cool air as needed (single duct). Power costs can be relatively high in constant flow systems because the fan runs continuously and this

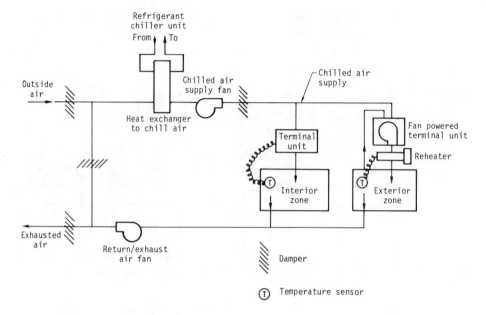

Figure 2.6 Single duct HVAC system with chilled air duct. Air is warmed when required by a reheater in the terminal unit or an in-room heater.

fixed air volume may require some cooling or heating during a major part of the day.

Variable Air Volume (VAV) systems attempt to save money by varying the volumetric airflow as well as air temperature. Sophisticated terminal units adjust air volume to control temperature; the system fan is equipped with inlet dampers (described in Chapter 9) to reduce power consumption when a lower air volume is required. The minimum air volume for each zone should be set to meet or exceed airflow criteria such as ASHRAE 62–1981 [11] but some studies have identified extremely low VAV unit airflows as the cause of problems.[4] There are at least two explanations for these extremely low VAV flow rates:

- Some HVAC systems were designed to energy conservation (that is, low airflow) criteria that experience has shown were overly stringent.
- In some buildings, occupant-generated heat in the interior zones migrates to the perimeter and so interior room temperature seldom rises enough to trigger the thermostat to call for more cool air from the HVAC system. The interior zone VAV terminals will thus operate at minimum airflow for most of the time and ventilation will be inadequate.[4]

Increasing the airflow to poorly ventilated interior zones may cause too great a temperature drop if the interior terminal devices lack reheaters.

Other Factors

In addition to the basic system design, the following factors should be considered:

- The quantity of outdoor air entering the HVAC system is usually adjustable (with a minimum preset level) to control temperature and relative humidity. Evaluations of air quality, particularly the buildup of contaminants, should consider the amount of outdoor air entering the building during episodes of air quality problems.
- The number and complexity of temperature and humidity sensors and computer controls vary widely in HVAC systems. Sophisticated control systems monitor conditions at many points and adjust outside air dampers, chiller and heater operation, and other parameters to minimize energy costs. Out-of-calibration sensors and control system malfunctions can increase energy consumption and also degrade indoor air quality by restricting airflow to some zones below design rates or by setting outside air dampers so that too little fresh air enters the system.
- Some terminal devices operate by drawing room air into the unit to recirculate it, thereby increasing mixing in the occupied space. Malfunctioning damper controls can impede adequate recirculation through these devices.

VENTILATION SOLUTIONS

This section covers ways to solve indoor air quality problems by changes to the ventilation system. It assumes that an evaluation of the problem has been completed and that thermal discomfort, lack of air movement, or presence of a contaminant amenable to removal by ventilation is the cause. Since these techniques are applied sequentially, it is important to maintain open communications with building occupants so that they will be patient while a solution is found.

HVAC System Adjustments
and Modifications

These steps involve adjustments or changes to the HVAC system:

- Where low air circulation or too high temperatures contribute to the problem, the easiest solution is lowering the thermostat setting by several degrees. For most HVAC systems this will increase volumetric airflow to the space as well as lower the temperature. For constant flow systems the air volume will not increase but the lower temperature will compensate for the apparent higher temperature caused by insufficient air motion. Lowering the thermostat setting will increase energy costs but at least one study showed the cost was not excessive with a modern HVAC system.[4]

- The next step is to review the system operating controls and settings. For example, systems may be set to operate only during periods when the building is occupied. It is also common practice to minimize the quantity of outdoor air drawn into the building during seasons when the air is heated or cooled. In cases like these, where buildup of contaminants can occur, extending fan system operating periods and increasing fresh air quantities to dilute odors and contaminants may help to solve the problem.
- The mechanical and system electrical components should be checked for proper operation and condition. For newer buildings it is possible that the system never functioned properly or that some components are overdue for routine cleaning or maintenance. Literature reports on tight building syndrome cite causes such as toilet exhaust fans that recirculated air to the building [4] and air filters that were grossly contaminated with microbial growth after several years of service without cleaning.[9] In older buildings the gradual development of mechanical problems, such as non-functioning dampers, can lead to problems. The typical items to review are listed in Table 2.5.
- If the air circulation rates are still too low after the previous steps

Table 2.5 Typical HVAC System Inspection Items

Component	Inspection Item
Fan and Drive[a]	Belt slippage and bearing wear. Fan inlet dampers, if present, for proper operation. Fan blades for dust buildup.
Ducts and Dampers	Outside air, return air and supply air dampers for free movement. Fire dampers—check that they are open.
General	Outside air intakes for debris buildup. Chiller, dehumidifier, and other internal surfaces for cleanliness. Filters for cleanliness, clogging, and scheduled replacement date.

[a]See Table 12.4 for safety precautions before inspecting fan.
Source: Reference 16.

have been applied, then the fan speed has to be increased, the fan replaced with a larger unit, or the terminal devices modified or replaced to increase air delivery and/or air motion in the occupied space. The best solution for each specific situation is based on final cost and depends on factors such as whether problems are limited to a few zones or involve the entire building. Fan modifications are discussed in Chapter 9 (Fans) and Chapter 13 (Solving Problems). Increasing the fan speed can be an expensive solution, since the energy used increases with the cube of the increase in airflow. For example, increasing airflow by 20 percent will increase fan power consumption by 73 percent, that is, $1.2 \times 1.2 \times 1.2 = 1.73$, just to move the air; extra heating or cooling energy to temper the additional air is added to this.

- When these steps do not work, a redesign of the control system or a major upgrading of the HVAC system may be needed, particularly if the original design was based on extreme energy conservation targets.

At each point in this step-by-step process the advantages of other control methods should be considered. Rather than perform some of the costly items described, it may be economically attractive to remove or seal sources of contaminants or to limit smoking.

Local Exhaust Systems

Use of local exhaust systems to remove contaminants before they are dispersed in the general building air should be considered. Some commercial equipment that produces vapors or dust can be connected to a duct system. Small rooms where photocopy machines are located can be exhausted to the outside so that vapors do not escape into the rest of the building. Use of spray coatings and fixatives used in graphic arts work can be restricted to an exhaust hood. Local exhaust can also remove heat generated by equipment.

Air Cleaners

The types and operating principles of air cleaners are described in Chapter 7. This section highlights factors applying to air cleaners used for indoor air pollution problems.

All HVAC systems already have one air cleaner—the filter in the return air ducts. This has a measurable effect on airborne particulate levels; when supply fans are run intermittently rather than continuously to save energy, the particulate levels rise. This is because the filtering action on resuspended dust is lost when the fan is off.[15]

When airborne particulates cause problems, more efficient filters or use of an electrostatic air cleaner may be beneficial:

- As the efficiency of a filter in removing small particulates increases, the resistance to airflow that it causes also increases. Often a standard furnace filter is used in front of the higher efficiency filter to capture the larger particles and thereby prolong its service life. As filters clog, the airflow decreases until they are cleaned or replaced.
- Electrostatic precipitators used in indoor air pollution applications are called "electronic air cleaners." They function by charging the particulates in an electric arc and then collecting them in an electric field. Electronic cleaners cause less resistance to airflow than highly efficient filters, but if not cleaned as required their collection efficiency declines.

If respirable particulate matter from tobacco smoking is too high, these particulate removing devices may be helpful. However, particulate devices alone may not solve the problem, since gaseous components may evolve from the captured particulates.[2] The gases must be collected in an activated carbon adsorber used along with the particulate device.

Gaseous air cleaners are also used for other organic vapor contaminants. The adsorber material is usually activated carbon or activated alumina impregnated with other chemicals such as potassium permanganate (an oxidizing agent) to control odors.

For all of these air cleaners, periodic replacement or maintenance is important.

Air Disinfection

Where transmission of disease is found to be caused by airborne pathogens, disinfection with ultraviolet (UV) radiation from mercury vapor discharge lamps has been used. The UV source is shielded to prevent direct exposure to occupants, house plants and furnishings, since UV radiation can cause skin and eye irritation and also fading of colors.[2] Generally the air above the heads of occupants is irradiated and normal mixing is relied upon to circulate pathogens into the irradiated area. If pathogens are transmitted through recirculating air ducts, UV lamps can be used inside the ducts to disinfect the air. Air disinfection is not routinely used in commercial and industrial buildings.

SUMMARY

The quality of indoor air is a complex issue. Problems have increased as buildings have become more tightly sealed and air exchange rates have been reduced. Often the health/safety professional or the HVAC engineer becomes involved in a situation only after it has reached crisis proportions.

Ventilation is still the primary way to remove most contaminants and provide adequate air quality and thermal conditions. Evaluation of the system along with air analyses to identify the contaminants and their levels will show whether the system is operating properly and whether design airflow rates are adequate to maintain a healthful and comfortable environment. Local exhaust systems and air cleaners are ventilation-related approaches that will solve some problems.

It is important to remember that some contaminants, such as asbestos, are usually not amenable to control by the HVAC system. Removal or another control technique may be required.

REFERENCES

1. Turk, A. "Gaseous Air Cleaning Can Help Maintain Tolerable Indoor Air Quality Limits," *ASHRAE J.* 25, No. 5, 35–7 (1983).
2. National Research Council-Committee on Indoor Pollutants. *Indoor Pollutants* (Washington, D.C.: National Academy Press, 1981).
3. Turiel, I., et al. "The Effects of Reduced Ventilation on Indoor Air Quality in an Office Building," *Atmospheric Environment* 17, No. 1, 51–64 (1983).
4. Carlton-Foss, J.A. "The Tight Building Syndrome," *ASHRAE J.* 25, No. 12, 38 (1983).
5. Yocum, J.E. "Indoor-Outdoor Air Quality Relationships: A Critical Review," *J. Air Pollution Control Assoc.* 32, No. 5, 500–20 (1982).
6. Hawthorne, A.R. and R.B. Gammage. "Formaldehyde Release from Simulated Wall Panels Insulated with Urea-Formaldehyde Foam Insulation," *J. Air Pollution Control Assoc.* 32, No. 11, 1126–31 (1982).
7. Cain, W.S., et al. "Ventilation Requirements in Buildings—I. Control of Occupancy Odor and Tobacco Smoke Odor," *Atmospheric Environment* 17, No. 6, 1183–97 (1983).
8. Spengler, J.D. and K. Sexton. "Indoor Air Pollution: A Public Health Perspective," *Science* 221, No. 4605, 9–17 (1983).
9. Bernstein, R.S., et al. "Exposures to Respirable Airborne *Penicillium* from a Contaminated Ventilation System: Clinical, Environmental and Epidemiological Aspects," *Amer. Industrial Hygiene Assoc. J.* 44, No. 3, 161–9 (1983).
10. Yaglou, C.P., E.C. Riley, and D.I. Coggins. "Ventilation Requirements," *ASHVE Trans.* 42, 133–162 (1936).
11. American Society of Heating, Refrigerating and Air Conditioning Engineers. "ASHRAE Standard 62–1981. Ventilation for Acceptable Indoor Air Quality." (New York, New York: ASHRAE, 1981).
12. American Society of Heating, Refrigerating and Air Conditioning Engineers. "ASHRAE Standard 90–75. Energy Conservation in New Building Design," (New York, New York: ASHRAE, 1977).
13. Turiel, I. and J. Rudy. "Occupant Generated CO_2 as an Indicator of Ven-

tilation Rate," UC-LBL Report No. LBL-10496 (Berkeley, California: Lawrence Berkeley Laboratory, 1980).

14. Tamblyn, R.T. "Beating the Blahs for VAV," *ASHRAE J.* 25, No. 9, 42–5 (1983).

15. Weschler, C.J., S.P. Kelty, and J.E. Lingousky. "The Effect of Building Fan Operation on Indoor-Outdoor Dust Relationships," *J. Air Pollution Control Assoc.* 33, No. 6, 624–6 (1983).

16. Whalen, J.M. "Two Techniques for Selling Energy Management," *Heating/Piping/Air Conditioning* 55, No. 9, 97–8 (1983).

CHAPTER 3

OSHA Ventilation Standards

The federal Occupational Safety and Health Act (OSHA) heightened interest in industrial ventilation. Practically overnight recommendations and guidelines published by the American National Standards Institute (ANSI), the National Fire Protection Association (NFPA), the American Conference of Governmental Industrial Hygienists (ACGIH), and others became law. The U.S. Department of Labor adopted these recommendations and guidelines, since they represented "consensus" standards and are part of the OSHA standards appearing in Chapter 26, Part 1910 of the U.S. *Code of Federal Regulations*.[1] Since states were encouraged to establish their own occupational safety and health programs under the federal OSHA law, many of the federal health standards are repeated in the individual state standards. However, the states are free to set their own standards as long as they are "at least as effective as" the federal standards. Regardless of whether or not your plant is covered by federal or state OSHA standards, be alert for changes in the requirements. The standards setting process will probably never be completed, since the odds are that new toxicological information will be developed about industrial chemicals and other substances.

One note of caution: this chapter covers only OSHA standards relating to ventilation. The same sections of the standards that mention ventilation may also set other requirements such as respirator use, labeling or posting signs, air monitoring to determine exposures, or maximum storage limits for flammable materials. Since this chapter is intended to help explain ventilation requirements, it does not cover the other standards that apply along with ventilation standards.

PURPOSE OF VENTILATION STANDARDS

Ventilation is required in OSHA standards for two reasons: to control employee exposures to potentially harmful materials and to prevent fire or explosion hazards. Within both categories there are two types of re-quirements: *performance* standards that set a goal (such as maintaining exposure to airborne contaminants below the OSHA permissible limit) and let the employer meet that goal in the best way; and *specification* standards that require the installation of a ventilation system for certain processes and a compliance with specific design requirements. An ex-ample of a specification standard is the requirement that whenever zinc-bearing or -coated metal is welded indoors, local exhaust ventilation be provided and the ventilation system move enough air to maintain an air velocity of 100 ft/min through the welding zone when working the maximum distance from the hood opening.[2] Specification standards do not give the employer much latitude, but they generally identify situ-ations that are recognized as hazardous enough to warrant controls. Some standards are combinations of performance and specification cri-teria. For example, when welding or torch cutting on cadmium-bearing or -coated metal is done indoors, either local ventilation or airline res-pirators are required "unless atmospheric tests under the most adverse conditions have established that the worker's exposure is within the acceptable concentrations" allowed by OSHA.[2]

The remainder of this chapter summarizes the federal OSHA stan-dards that mention ventilation. These standards are discussed either under "health" or "fire and explosion prevention" categories. The stan-dards for some processes and materials have provisions for both health and fire/explosion prevention. These standards are discussed under the category that applies predominantly. Specific substances mentioned in this chapter include asbestos, 13 chemicals classed by OSHA as carcin-ogens; flammable or combustible liquids; hydrogen and oxygen. Specific operations include abrasive blasting; grinding, polishing, and buffing; open surface tanks; welding, cutting, and brazing; and spray finishing.

HEALTH STANDARDS

The aim of health ventilation standards is to prevent exposures that exceed the limits allowed by OSHA chemical exposure standards. There are general exposure standards that set permissible exposures to chem-icals, and standards that cover specific operations.

General Exposure Standards

The general exposure standards in Part 1910.1000 of the *Code of Federal Regulations* define the allowable exposure to specific chemicals. For most substances this allowable exposure is expressed as the average exposure over an eight-hour shift. For some substances, however, a ceiling (maximum) exposure limit is also specified along with an allowable "peak" short-term exposure above the ceiling. The unit of exposure for vapors and gases is usually parts per million (ppm), meaning parts of contaminant per million parts of contaminated air (volume:volume). For dusts, fumes, and other particulates the unit of measurement is usually milligrams of contaminant per cubic meter of air (mg/m^3), although asbestos fiber concentrations are expressed as asbestos fibers per cubic centimeter of air. Figure 3.1 shows examples of the different types of standards.

The time-weighted average exposure standards recognize that employee exposures vary over the work day and that the toxicological effect of many substances depends on the average exposure over the whole day. For other substances the average exposure standard concept is not valid. The substance is either toxic or irritating for short exposures to sufficiently high concentrations. Some states have adopted excursion factors from the ACGIH TLV booklet along with the general TWA exposure standards.[3] The excursion factors set automatic ceiling limits on exposures depending on the magnitude of the allowable time-weighted average exposure.

The time-weighted average (TWA) exposure, also called the cumulative exposure in OSHA standards, is calculated from this equation:

$$\text{TWA} = \frac{(C_1 \times T_1) + (C_2 \times T_2) + (C_3 \times T_3) + \ldots + (C_n \times T_n)}{8} \quad (3.1)$$

where TWA = time-weighted average exposure, same units as C
$C_1, \ldots C_n$ = concentration of substance during different time
periods throughout the day, usually ppm or mg/m^3
$T_1, \ldots T_n$ = time duration of the exposure at the concentration
C, hr

If the calculated TWA does not exceed the standard, no legal violation exists. Of course any exposure that causes apparent adverse health effects should be reduced even if the exposure meets legal standards.

Example: Workers in a plant manufacturing rayon staple fiber by the viscose process are exposed to carbon disulfide (CS_2).[4] Breathing zone air samples collected on one worker are:

Time Duration	CS_2 Concentration
2 hr	2.1 ppm
2 hr	22.0 ppm
3 hr	6.0 ppm
1 hr	3.9 ppm
2-min peak	85 ppm

The exposure pattern is illustrated in Figure 3.2. Calculate the time-weighted average exposure and compare the exposure to OSHA permissible limits for carbon disulfide in Figure 3.1.

Substance	p/m[a]	mg./M[b]
2-Hexanone	100	410
Hexone (Methyl isobutyl ketone).......	100	410
sec-Hexyl acetate	50	300
Hydrazine—Skin	1	1.3
Hydrogen bromide	3	10
C Hydrogen chloride...........................	5	7
Hydrogen cyanide—Skin.....................	10	11
Hydrogen peroxide (90%)...................	1	1.4
Hydrogen selenide	0.05	0.2
Hydroquinone		2
C Iodine...	0.1	1
Iron oxide fume		10
Isoamyl acetate.................................	100	525
Isoamyl alcohol	100	360
Isobutyl acetate.................................	150	700
Isobutyl alcohol	100	300
Isophorone...	25	140
Isopropyl acetate...............................	250	950
Isopropyl alcohol...............................	400	980
Isopropylamine	5	12
Isopropylether....................................	500	2,100

[a] Parts of vapor or gas per million parts of contaminated air by volume at 25° C. and 760 mm. Hg pressure.
[b] Approximate milligrams of particulate per cubic meter of air.

Material	8-hour time weighted average	Acceptable ceiling concentration	Acceptable maximum peak above the acceptable ceiling concentration for an 8-hour shift.	
			Concentration	Maximum duration
Beryllium and beryllium compounds (Z37.29–1970)	2 μg./M^3............	5 μg./M^3............	25 μg./M^3	30 minutes.
Cadmium fume (Z37.5–1970)	0.1 mg./M^3	0.3 mg./M^3.......	....	
Cadmium dust (Z37.5–1970)	0.2 mg./M^3	0.6 mg./M^3...... .	....	
Carbon disulfide (Z37.3–1968)	20 p.p.m............	30 p.p.m............	100 p.p.m	30 minutes.
Carbon tetrachloride (Z37.17–1967)	10 p.p.m............	25 p.p.m............	200 p.p.m	5 minutes in any 4 hours.

Figure 3.1 Examples of federal OSHA exposure standards: *(Top)* Time-weighted average and ceiling exposure standards. The "C" preceding the substance name (e.g., Iodine) indicates a ceiling value. *(Bottom)* Time-weighted average plus ceiling and peak above ceiling exposure standards.

Answer: Using Equation 3.1 the time-weighted average exposure is:

$$TWA = \frac{(2.1 \text{ ppm}) 2 + (22.0 \text{ ppm}) 2 + (6.0 \text{ ppm}) 3 + (3.9 \text{ ppm}) 1}{8}$$

$$TWA = \frac{4.2 + 44.0 + 18.0 + 3.9}{8} = 8.8 \text{ ppm}$$

According to the federal OSHA exposure standards in Figure 3.1, the 8.8 ppm TWA exposure is below the 20 ppm permitted, and the ceiling exposure of 22.0 ppm is below

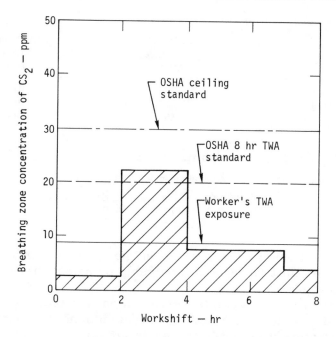

Figure 3.2 Worker's exposure pattern to carbon disulfide, showing OSHA exposure standards, used in example.

the allowed ceiling of 30 ppm. The 2-min peak sample of 85 ppm is also below the 100-ppm permissible peak. So this exposure does not exceed OSHA standards.

If an employee is exposed to a mixture of contaminants, the employer must calculate the equivalent exposure by summing the contribution of each substance to the overall exposure:

$$E = \frac{C_1}{L_1} + \frac{C_2}{L_2} + \ldots + \frac{C_n}{L_n} \qquad (3.2)$$

where E = equivalent exposure for the mixture, dimensionless
$C_1, \ldots C_n$ = concentration of a particular contaminant, usually ppm or mg/m^3
$L_1, \ldots L_n$ = exposure limit for that contaminant, same units as C

If the calculated equivalent exposure does not exceed one, an overexposure does not exist according to OSHA standards.

Example: A worker is exposed to isoamyl alcohol, isobutyl alcohol, and isopropyl alcohol during the same day. Air samples show the time-weighted average exposure to each, along with the OSHA permissible exposure from Figure 3.1, to be:

Substance	Worker's TWA Exposure (ppm)	OSHA Permissible 8-hr Exposure (ppm)
Isoamyl alcohol	25	100
Isobutyl alcohol	15	100
Isopropyl alcohol	150	400

Calculate the equivalent exposure to this mixture.

Answer: From Equation 3.2 the equivalent exposure is:

$$E = \frac{25 \text{ ppm}}{100 \text{ ppm}} + \frac{15 \text{ ppm}}{100 \text{ ppm}} + \frac{150 \text{ ppm}}{400 \text{ ppm}} = 0.78$$

Since E is less than 1.0, no overexposure exists.

If overexposures exist, either to single substances or to mixtures, the exposure must be reduced, preferably by administrative or engineering controls.

In most cases an accurate determination of employee exposures is the first step in deciding whether ventilation is needed to meet these general exposure criteria. Chapter 4 (Hazard Assessment) outlines how to determine employee exposures.

There are several specific health standards, such as Lead (1910.1025) and Cotton Dust (1910.1043), that do not require ventilation. They state, however, that when ventilation is used to control exposures the system must be tested periodically or within a short time of any change in production, process, or control that could affect exposures. Typical tests that measure the system's effectiveness include capture velocity, duct velocity, and hood static pressure as discussed in Chapter 12.

Exposure Standards Requiring Ventilation

Some federal OSHA contaminant exposure standards specifically mention ventilation. For example, the standard governing exposure to asbestos (Part 1910.1001) specifies that hand-operated and power-operated tools which may release or produce asbestos fibers in excess of the legal limit must be provided with local exhaust ventilation. Also, dusty asbestos-containing materials must be enclosed, wetted, or ventilated when removed from the original container.

Federal OSHA standards also contain standards for 13 compounds identified as suspected or known carcinogens (cancer-causing agents). These substances are listed in Table 3.1. The standards for all 13 substances are essentially identical and are designed to minimize potential exposure to these materials. The standards speak of "isolated systems,"

Table 3.1 Standards For 13 OSHA Carcinogens

Substance	OSHA Standard
4-Nitrobiphenyl	1910.1003
alpha-Naphthylamine	1910.1004
Methyl chloromethyl ether	1910.1006
3,3'-Dichlorobenzidine (and its salts)	1910.1007
bis-Chloromethyl ether	1910.1008
beta-Naphthylamine	1910.1009
Benzidine	1910.1010
4-Aminodiphenyl	1910.1011
Ethyleneimine	1910.1012
beta-Propiolactone	1910.1013
2-Acetylaminofluorene	1910.1014
4-Dimethylaminoazobenzene	1910.1015
N-Nitrosodimethylamine	1910.1016

such as glove boxes, with absolute filters on the exhaust system. They also require that ventilation be provided to produce air movement from the work area to the contaminant source whenever these materials are removed from closed systems. Also, all exhausted air must be decontaminated before discharge and the correct volume of make-up or replacement air must be supplied to the workroom to keep the exhaust system working properly.

For operations with these 13 materials involving laboratory-type hoods (Figure 3.3), an average velocity of 150 ft/min through the hood openings with a minimum velocity of 125 ft/min at any point is required. Laboratory hoods or glove boxes are required for operations or equipment that could generate aerosols. In order to ensure that there is no leakage from the contaminated area, all areas in which these substances are handled must have lower air pressure than surrounding work areas and the outside environment.

In laboratories and animal rooms in which these 13 materials are used, special requirements apply. There can be no connection with other areas of the building through the ventilation system. All ventilated apparatus, such as hoods, must be tested semiannually.

Ventilation Standards for Health Protection

Part 1910.94 in the federal OSHA standards covers ventilation requirements for different industrial operations. Since the wording in this section is taken from both ANSI and NFPA standards, the requirements vary from one operation to another. For example, some standards require periodic testing of ventilation systems while others do not. In general,

Figure 3.3 Laboratory hood showing hood face velocity measurements. *(Courtesy Kewaunee Scientific Equipment Corp.)*

however, the intent is to keep exposures below the permissible levels described earlier, either by ventilation or some other control technique.

Abrasive Blasting

Abrasive blasting is the cleaning of a surface using a stream of abrasive applied by pressure or force. The potential hazard is from the dust as abrasive is pulverized and from the materials removed from the surface being cleaned. Silica exposure from sandblasting is a well-known hazard from abrasive cleaning operations. Some dusts are combustible and hazardous if inhaled. For these operations all equipment, including the ventilation system, must meet specific standards. Combustible organic abrasives can only be used in automatic abrasive blasting systems.

   The federal OSHA standards for abrasive blasting require that concentration of respirable dust or fumes in the workers' breathing zone not exceed the exposure limits for the specific materials involved. This implies that you know what the dust composition is and whether the levels are hazardous.

   Abrasive blasting is done either inside an enclosure (rotating barrel, blasting cabinet, or walk-in room) or in the open. Abrasive blasting respirators (Figure 3.4) are required for personnel working inside a blast-

Figure 3.4 An air-supplied abrasive blasting respirator and protective hood.

ing room or doing manual sandblasting when the nozzle and blast are not separated from the worker inside a ventilated enclosure. The respirators are also used for any other operation where the workers' exposure could exceed allowable limits during blasting. Dust mask respirators are required during cleanup or other tasks where ventilation is not feasible and dust concentrations can exceed standards. A respirator program that meets OSHA requirements in Part 1910.134 is needed whenever respirators are used for health protection.

When abrasive blasting is conducted inside an enclosure, it must be ventilated to obtain a continuous inward flow of air through openings during blasting. Openings must be baffled or located to minimize the escape of abrasives and dust and there may be no visible spurts of dust. Also, the ventilation system must clear the enclosure of dust promptly after blasting ceases.

The construction, installation, inspection, and maintenance of the exhaust system must follow ANSI standards Z 9.2–1960 [5] and Z 33.1–1961. [6] Specific items from these standards incorporated into OSHA abrasive blasting requirements are: repair of dust leaks; periodic measurement of static pressure to determine blockages; the use of a fines separator when abrasive is recirculated; and a dust collector in the exhaust system that can be emptied of accumulated dust without creating a hazard.

Grinding, Polishing, and Buffing

The federal OSHA standards for grinding, polishing, and buffing (GPB) in Part 1910.94b contain some of the most detailed ventilation requirements in OSHA. Local exhaust ventilation is required when "dry grinding, dry polishing, or buffing is performed, and employee exposure, without regard to the use of respirators, exceeds the permissible exposure limits."

There are numerous figures specifying hood design and minimum airflow rates (Figure 3.5) for different types of hoods, generally taken from ANSI Z 43.1–1966 "Ventilation Control of Grinding, Polishing and Buffing Operations"[7] and other ANSI standards.[8]

This section of the standards illustrates how the standards-setting process can be carried too far. OSHA originally adopted "consensus" standards by reviewing ANSI publications and other standards and changing words like *should* to *shall* to produce a legally enforceable standard. This can result in advisory statements being made part of a mandatory standard. For example, the OSHA standard for GPB hoods states that "the entry loss from all hoods except the vertical spindle disc grinder shall equal 0.65 velocity pressure . . ." The hood entry loss factor is the resistance to airflow into the hood caused by turbulence and friction. It is reduced by design features that minimize friction and turbulence. A value for hood entry loss is assumed during system design so the ducts and fan that draw the required airflow can be calculated as explained in Chapter 5. These values are listed in design tables based on experimentation and experience. However, there is no way to design

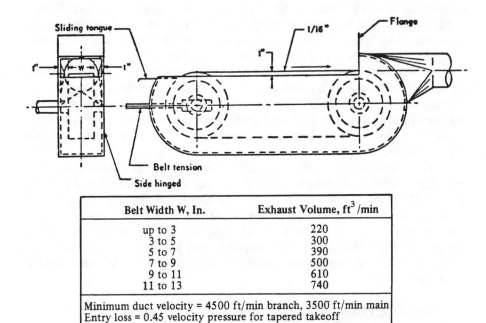

Belt Width W, In.	Exhaust Volume, ft^3/min
up to 3	220
3 to 5	300
5 to 7	390
7 to 9	500
9 to 11	610
11 to 13	740
Minimum duct velocity = 4500 ft/min branch, 3500 ft/min main Entry loss = 0.45 velocity pressure for tapered takeoff 0.65 velocity pressure for straight takeoff	

Figure 3.5 Belt grinder hood design from federal OSHA standards.

a hood to give a predetermined hood entry loss once it is built and in service. If the hood draws the required amount of air, the system meets OSHA standards; the hood entry loss has no bearing on airflow through the hood.

If you are designing hoods or enclosures for these operations it is a good idea to follow the design figures in OSHA (Figure 3.5) or similar ones in the ACGIH *Industrial Ventilation Manual*[8], if practical. They represent designs that work based on experience. However, a study sponsored by the National Institute for Occupational Safety and Health (NIOSH) showed that standard hood and enclosure designs may not be adequate in reducing exposures when high-toxicity materials are used.[10] Enough contaminants may escape from the operation to cause employee overexposures. Special precautions for ventilation systems handling high-toxicity or high-nuisance substances are covered in Chapter 10.

Open Surface Tanks

Part 1910.94d covers all operations involving immersion of materials in liquids or the vapors of liquids contained in open surface tanks for cleaning or other surface alterations. Examples include vapor degreasing, electroplating, pickling, dyeing, and bleaching. Ventilation is only one form of control that may be used to minimize the hazards of open surface tanks. For example, tank covers, foams, beads, or other surface tension depressive agents may be used along with ventilation. Ventilation requirements are found according to a prescribed series of steps that start with a Hazard Classification of the operation. The classification is based on toxicity, potential for fire, and method of use of the materials.

To design ventilation for open surface tanks, classify the material into one of 18 classes A–1 through D–4 according to this procedure:

* Determine the Hazard Potential using the OSHA allowable concentration of contaminants from Part 1910.1000 *et seq.* or their flash point. For mixtures, except organic solvents, determine the Hazard Potential from the most hazardous component that constitutes a significant fraction of the mixture. For organic solvents calculate the combined exposure from Equation 3.2. The criteria for letter ratings "A" through "D" for Hazard Potential are found in Figure 3.6.
* Determine the rate of gas, vapor, or mist evolution on a scale of 1–4 depending on the following factors: how far the liquid is below its boiling point; what the relative evaporation rate is in still air at room temperature on an arbitrary scale of fast, medium, slow, or nil; and how much mist or gas the tank produces, on the same arbitrary scale. Use Figure 3.7 to determine the rate of evolution value.

| Hazard Potential | Toxicity Group | | |
	Gas or Vapor ppm	Mist mg/m^3	Flash Point °F
A	0-10	0-0.1	
B	11-100	0.11-1.0	Under 100
C	101-500	1.1-10	100-200
D	Over 500	Over 10	Over 200

Figure 3.6   Federal OSHA criteria for determining the hazard potential for open surface tank operations.

Examples of operations falling into the different hazard classes are listed in Table 3.2. For your convenience the ACGIH *Industrial Ventilation Manual*[9] lists the classifications of many common open surface tank operations.

Rate	Liquid Temperature °F	Degrees below Boiling Point	Relative Evaporation[a]	Gassing[b]
1	Over 200	0-20	Fast	High
2	150-200	21-50	Medium	Medium
3	94-149	51-100	Slow	Low
4	Under 94	Over 100	Nil	Nil

Note: In certain classes of equipment, specifically vapor degreasers, an internal condenser or vapor level thermostat is used to prevent the vapor from leaving the tank during normal operation. In such cases, rate of vapor evolution from the tank into the workroom is not dependent upon the factors listed in the table, but rather upon abnormalities of operating procedure, such as carryout of vapors from excessively fast action, dragout of liquid by entrainment in parts, contamination of solvent by water and other materials or improper heat balance. When operating procedure is excellent, effective rate of evolution may be taken as 4. When operating procedure is average, the effective rate of evolution may be taken as 3. When operation is poor, a rate of 2 or 1 is indicated, depending upon observed conditions.

[a]Relative evaporation rate is determined according to the methods described by A. K. Doolittle in *Industrial and Engineering Chemistry*, vol. 27, p. 1169, where time for 100% evaporation is as follows: Fast, 0-3 hr; Medium, 3-12 hr; Slow, 12-50 hr; Nil, more than 50 hr.

[b]Gassing means the formation by chemical or electrochemical action of minute bubbles of gas under the surface of the liquid in the tank and is generally limited to aqueous solutions.

Figure 3.7   Federal OSHA criteria for determining the rate of evolution value for open surface tank operations.

Table 3.2 Typical OSHA Hazard Classifications For Open Surface Tank Operations

Classification	Process	Temperature Range, °F	Contaminants
A-1	Chromium electroplating in chromic acid bath	90–140	Chromic acid mist
A-2	Etching copper in hydrochloric acid bath	70–90	Hydrogen chloride gas
A-3	Pickling aluminum in chromic and sulfuric acid bath	140	Acid mists
A-4	Zinc dip in chromic and hydrochloric acid bath	70	Hydrogen chloride gas
B-1	Pickling iron and steel in sulfuric acid bath	70–175	Sulfuric acid mist, steam
B-2	Nickel electroplating in nickel sulfate bath	70–90	Nickel sulfate mist
B-3	Zinc electroplating in zinc chloride bath	75–120	Zinc chloride mist
B-4	Stripping tin in ferric chloride, copper sulfate and acetic acid bath	70	Acid mist
C-1	Etching aluminum in sodium hydroxide, soda ash, and trisodium phosphate bath	160–180	Alkaline mist, steam
C-2	Copper strike in cyanide bath	70–90	Cyanide mist
C-3	Tin electroplating in sodium stannate bath	140–170	Tin salt mist, steam
C-4	Gold electroplating in cyanide bath	75–100	Cyanide mist
D-1	Magnesium dye set	212	Steam
D-2	Parkerizing	140–212	Steam
D-3	Hot water baths	100–200	Water vapor
D-4	Copper etching in ferric chloride bath	70	None

Source: ACGIH *Manual*.[9]

Once you determine the hazard class, choose one of the following types of hoods: enclosure, lateral exhaust, or canopy hood. (These hood types are described in Chapter 6.) Then use Figure 3.8 to find the needed control velocity for the hazard class and hood type. The control velocities apply to cases where airflow is undisturbed by local environmental conditions such as open windows, wall fans, unit heaters, or moving machinery.

With this information find the required minimum volumetric flow rate (ft³/min) through the hood:

- For enclosures the airflow rate is the control velocity multiplied by the open areas through which air flows into the enclosure.
- For canopies the airflow rate is the control velocity multiplied by the net area of all openings between the bottom edges of the hood and the top edges of the tank. Adding one or more sides to a canopy to reduce the open area significantly reduces exhaust requirements.
- For lateral hoods the minimum ventilation rate expressed as ft³/min per ft² of tank surface area is found in Figure 3.9 based on the tank width-to-length ratio assuming the hood is along the tank length. If the tank has hoods along two opposite sides, consider the distance each hood must reach as the tank width for this calculation (Figure 3.10). Multiply the ventilation rate from Figure 3.9 by the tank surface area to find the volumetric airflow for the hood. Note that these ventilation rates are the minimum for locations that are undisturbed by crossdrafts. Crossdrafts can reduce lateral hood efficiencies by up to 70% as described in Chapter 6. For lateral hood reaches exceeding 42 inches or for reduction of exhaust rates in smaller lateral hoods, a push-pull system may be used. A push-pull system (Figure 3.11) uses an air supply manifold on the tank edge opposite the exhaust

	Enclosing Hood			Canopy Hood[b]	
Class	One Open Side	Two Open Sides	Lateral Exhaust[a]	Three Open Sides	Four Open Sides
A-1 and A-2	100	150	150	Do not use	Do not use
A-3[b], B-1, B-2 and C-1	75	100	100	125	175
B-3, C-2 and D-1[c]	65	90	75	100	150
A-4,[b] C-3 and D-2[c]	50	75	50	75	125
B-4, C-4, D-3[c] and D-4	General room ventilation required.				

[a]See Figure 2-9 for computation of ventilation rate.
[b]Do not use canopy hood for Hazard Potential A processes.
[c]Where complete control of hot water is desired, design as next highest class.

Figure 3.8 Federal OSHA control velocity requirements in feet/minute for undisturbed locations for different classes of open surface tank operations.

Required Minimum Control Velocity, fpm (from Figure 2-8)	Cfm/ft^2 to Maintain Required Minimum Velocities at Following Ratios tank width (W)/tank length (L)[a,b]				
	0.0–0.09	0.1–0.24	0.25–0.49	0.5–0.99	1.0–2.0
Hood along one side or two parallel sides of tank when one hood is against a wall or baffle[b] Also for a manifold along tank centerline.[c]					
50	50	60	75	90	100
75	75	90	110	130	150
100	100	125	150	175	200
150	150	190	225	260	300
Hood along one side or two parallel sides of free standing tank not against wall or baffle.					
50	75	90	100	110	125
75	110	130	150	170	190
100	150	175	200	225	250
150	225	260	300	340	375

[a]It is not practicable to ventilate across the long dimension of a tank whose ratio W/L exceeds 2.0. It is undesirable to do so when W/L exceeds 1.0. For circular tanks with lateral exhaust along up to 1/2 the circumference, use W/L = 1.0; for over one-half the circumference use W/L = 0.5.

[b]Baffle is a vertical plate the same length as the tank, and with the top of the plate as high as the tank is wide. If the exhaust hood is on the side of a tank against a building wall or close to it, it is perfectly baffled.

[c]Use W/2 as tank width in computing when manifold is along centerline, or when hoods are used on two parallel sides of a tank. Tank width (W) means the effective width over which the hood must pull air to operate (for example, where the hood face is set back from the edge of the tank, this set back must be added in measuring tank width). The surface area of tanks can frequently be reduced and better control obtained (particularly on conveyorized systems) by using covers extending from the upper edges of the slots toward the center of the tank.

Figure 3.9 Federal OSHA lateral exhaust hood minimum airflow rates. Exhaust rate depends on the required control velocity from Figure 3.8 and the tank dimensions.

hood. Air is blown across the tank surface and into the exhaust hood. Consult the OSHA standard, the ACGIH *Manual,* and Chapter 6 of this book for the design of push-pull systems.

Example: A platinum electroplating bath (open surface tank) that is 3 ft long by 2 ft wide contains a solution of ammonium phosphate in water at 175°F. The tank is in the center of a room with no walls or baffles adjacent to it. Ammonia gas is released from the bath. Compare the exhaust flow rate for a lateral hood located at the tank edge with the rate for a canopy hood located 3 ft above the tank (Figure 3.12).

Answer: The OSHA PEL for ammonia is 25 ppm. Using Figure 3.6 the Hazard Potential for this operation is "B".

From Figure 3.7 the Rate of Gas Evolution is 2, since the bath is between 150°–200°F.

From Figure 3.8 the control velocity for B-2 processes is 100 ft/min for a lateral hood and 175 ft/min for a canopy hood with four sides open.

Figure 3.10 The tank width (W) used to calculate airflow rates is the distance over which the hood draws contaminants.

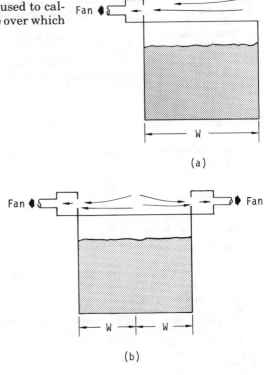

(a)

(b)

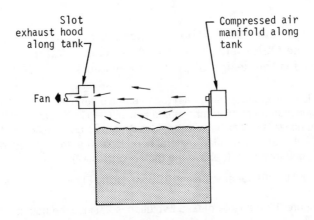

Figure 3.11 Push-pull systems are used on wide tanks where the width exceeds the capability of exhaust hoods to draw contaminants. Nozzles in the compressed air manifold blow air across the tank toward the exhaust hood.

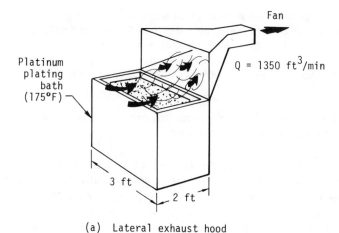

(a) Lateral exhaust hood

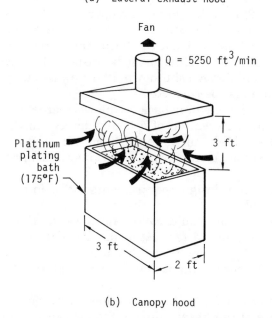

(b) Canopy hood

Figure 3.12 Airflow rates required by federal OSHA standards vary for different hood types. The canopy hood requires about four times more air than the lateral, or capturing, hood.

For the lateral hood the airflow is calculated from Figure 3.9. The tank width:length ratio is

$$\frac{2}{3} = 0.67$$

For a free-standing tank requiring 100 ft/min control velocity and with a width:length ratio of 0.67 the required airflow is 225 ft³/min per ft² of tank surface:

$$Q_{lateral} = 225 \, \frac{ft^3/min}{ft^2} \times 6 \, ft^2 = 1350 \, ft^3/min$$

For a canopy the open area is a four-sided box with dimensions 3 ft long × 2 ft wide × 3 ft high, which is the open area between the tank and canopy:

$$\text{Area} = (3 \times 3) + (3 \times 3) + (2 \times 3) + (2 \times 3)$$
$$= 9 + 9 + 6 + 6 = 30 \text{ ft}^2$$

The required airflow is the control velocity times the open area:

$$Q_{canopy} = 175 \frac{\text{ft}}{\text{min}} \times 30 \text{ ft}^2 = 5250 \text{ ft}^3/\text{min}$$

Therefore, to meet federal OSHA standards, the canopy hood requires almost four times as much airflow as the lateral exhaust hood.

If the operation being ventilated produces airborne spray, these droplets must be controlled to prevent escape into the workroom. Since it requires more airflow to capture droplets rather than gas or mist, enclosing the process as much as possible will reduce airflow requirements. Spray cleaning is a typical operation that generates spray droplets.

Follow ANSI Z 9.2[5] in designing and operating the exhaust system. For new systems the airflow must be adjusted until the proper flow through each hood is achieved. Then the hood static pressure must be measured and recorded. The system must be inspected at least every three months or after a prolonged shutdown. If airflow through any hood is too low, the system must be corrected.

Recirculation of exhausted air back to the work area is acceptable if air cleaners prevent the creation of a health hazard in the area receiving recirculated air. Make-up or replacement air, whether fresh or recirculated, must be supplied and the make-up air system must be tested periodically.

This summarizes the design requirements for open surface tank ventilation. Before tanks are cleaned, the liquid must be drained and pits where hazardous vapors could collect must be ventilated. If employees are to enter the tank, tests must be made to determine that the atmosphere is safe. If tests show that a hazardous environment exists from oxygen deficiency, contaminant level, or fire/explosion hazard, the tank must be ventilated to remove the hazardous condition before employees are allowed to enter the tank. Ventilation must then be continued to prevent recurrence of the hazard during maintenance. Welding and other operations that produce toxic metal fumes require a local exhaust or respirators.

Welding, Cutting, and Brazing

OSHA ventilation standards in Part 1910.252 cover both fire/explosion and health protection. For explosion prevention the storage of gases and carbide is regulated. Compressed gas cylinders must be stored in well-ventilated areas at least 20 ft from highly combustible materials. Cylinders shall not be stored in unventilated enclosures such as lockers and cupboards. Calcium carbide for acetylene generation must be stored in dry, waterproof, and well-ventilated locations. Depending on the amount of calcium carbide stored, different requirements (such as other occupancy in the buiding) exist, but all inside generator rooms or outside generator houses must be well ventilated with vents at floor and ceiling levels.

Ventilation health standards for these operations recognize three factors that govern welder exposure to contaminants: dimensions of the welding space, especially ceiling height; the number of welders; and possible evolution of hazardous fumes or dust according to the metals, fluxes, and other materials involved. Other factors, such as weather, the amount of heat generated, and the presence of volatile solvents, also contribute to the need for ventilation or respiratory protective devices. Welding screens, which shield nearby workers and onlookers from the ultraviolet light generated by the welding arc, should not restrict ventilation. They should be mounted about 2 ft above the floor unless a lower level is needed to protect others nearby.

Ventilation for welding and cutting of metals not specifically covered later in this section of standards is required under these conditions:

- In a space of less than 10,000 ft^3 per welder.
- In a room with ceiling height less than 16 ft.
- In confined spaces or where the area contains partitions or other structures that significantly obstruct cross-ventilation.

When ventilation is required for general welding and cutting, you can choose between general or local ventilation. For general ventilation either a minimum of 2000 ft^3/min per welder or an approved respirator is needed. For local exhaust ventilation either a moveable hood or a fixed enclosure is acceptable. An enclosure must develop an air velocity of at least 100 ft/min away from the welder. A movable hood must develop at least 100 ft/min in the welding zone. Table 3.3 summarizes these requirements. The OSHA standard lists the volumetric airflow needed to develop this velocity at varying distances from the hood opening. However, the airflow rates in the OSHA standards are lower than current recommendations in the ACGIH *Industrial Ventilation Manual*. The two standards' rates are compared in Figure 3.13. The important thing to remember is that achieving a specific volumetric airflow rate (ft^3/min) into the hood is not the goal of the standard; the 100-ft/min velocity in the welding zone is what reduces airborne contaminant levels.

General welding or brazing in confined spaces requires ventilation adequate to prevent accumulation of toxic materials or possible oxygen deficiency. This applies to helpers and other workers in the area as well as to the welder. The air that is exhausted from the confined space must be replaced with clean, respirable air. If ventilation cannot be provided, then approved airline respirators or hose masks must be used. In areas

Table 3.3 Federal OSHA Local Exhaust Requirements for General Welding and Cutting

Freely Movable Hood

Requires 100 ft/min velocity toward hood through welding zone when the hood is at its most remote point from the point of welding.

The airflow (ft³/min) needed to accomplish this velocity using a 3-inch-wide flanged suction opening is:

Welding Zone (distance from arc or torch), inches	Minimum Airflow,[a] ft³/min	Duct Diameter,[b] inches
4–6	150	3
6–8	275	3.5
8–10	425	4.5
10–12	600	5.5

Fixed Enclosure

Requires 100 ft/min velocity away from the welder.

A fixed enclosure, by definition, has a top and at least two sides which surround the welding or cutting operations.

[a] When brazing with cadmium-bearing materials or when cutting on such materials, increased rates of ventilation may be required.
[b] Nearest half-inch duct diameter based on 4,000 feet per minute velocity in pipe.

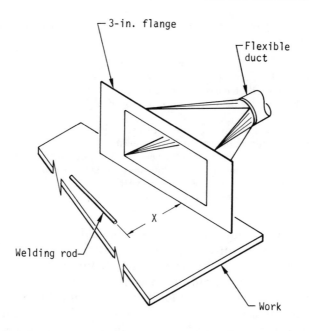

Welding zone X, in.	Airflow into hood, ft^3/min	
	Federal OSHA standards	ACGIH guidelines
4-6	150	250
6-8 or 9	275	560
8-10	425	NA
9-12	NA	1000
10-12	600	NA

Figure 3.13 Comparison of federal OSHA standards for movable welding hoods with the current (1982) ACGIH *Manual* guidelines.

with atmospheres that are immediately hazardous to life, a self-contained breathing apparatus is generally required and an outside helper equipped for rescue must be stationed to insure the safety of those working within the confined space. Oxygen must not be used to ventilate confined spaces. Requirements for general welding in confined spaces are listed in Table 3.4.

Operations involving specific toxic metals or other substances require local ventilation or approved respiratory protection without regard to workroom size and other conditions that govern whether ventilation is needed for general welding and cutting. OSHA welding standards cover these materials:

Table 3.4   Federal OSHA Ventilation Standards for General Welding and Cutting in Confined Spaces

---

Adequate ventilation is required to prevent accumulation of toxic materials or oxygen deficiency for welders, helpers, and other nearby personnel.

All replacement air shall be clean and respirable.

If it is impossible to provide ventilation, approved airline masks or hose masks are required.

In areas that are immediately hazardous to life, approved hose masks with blower or self-contained breathing apparatus shall be used.

If welders and helpers are provided with hose masks, hose masks with blowers or self-contained breathing apparatus, a worker shall be stationed outside of the confined space to ensure the safety of those working within.

Oxygen shall never be used for ventilation.

- Operations involving fluorine (as a chemical compound, not a gas) in confined spaces require adequate ventilation to prevent the accumulation of toxic material or oxygen deficiency. The same standards as those for general welding in confined spaces apply. Outside of confined spaces the need for local exhaust ventilation or airline respirators depends on contaminant levels, although "experience has shown such protection to be desirable for fixed-location production welding and for all production welding on stainless steels."[2]

- Welding or cutting involving zinc in confined spaces must meet the same requirements as general welding in confined spaces (Table 3.4). Indoors these operations require local exhaust booths or hoods meeting the standards in Table 3.3.

- Welding or cutting involving lead-base metals in confined spaces requires local exhaust ventilation meeting the requirements in Table 3.4. When these operations are conducted indoors, local ventilation as in Table 3.3 or airline respirators are required. Outdoors approved respirators are needed. In all cases workers in the immediate vicinity of the operation should be protected as necessary by local exhaust ventilation or airline respirators.

- Welding or cutting involving beryllium-containing base or filler metals whether conducted indoors, outdoors, or in confined spaces requires local exhaust ventilation *and* airline respirators. These standards apply unless air sampling under the most adverse conditions shows that beryllium exposures are below acceptable OSHA limits. As with lead, nearby workers may require protection by ventilation or airline respirators.

- Welding or cutting involving cadmium-bearing or cadmium-coated base metals indoors or in confined spaces requires local exhaust ventilation or airline respirators unless air sampling under the most adverse conditions shows that worker exposures are within acceptable limits. Outdoors such operations require approved respiratory protection such as fume respirators.

Welding (brazing) involving cadmium-bearing filler metals in confined spaces requires ventilation meeting the requirements in Tables 3.3 or 3.4.

- Welding or cutting indoors or in confined spaces involving metals coated with mercury-bearing materials (including paint) requires local exhaust ventilation or airline respirators unless air sampling under the most adverse conditions shows that mercury exposures are within acceptable limits. Outdoors such operations require approved respiratory protection.
- Cutting of stainless steels with oxygen using either a chemical flux or iron powder, or using a gas-shielded arc, requires mechanical ventilation adequate to remove the fumes generated.
- Degreasing or other cleaning operations involving chlorinated hydrocarbons must be located to prevent solvent vapors from reaching any welding operation. In addition, trichloroethylene and perchloroethylene should be kept out of atmospheres penetrated by the ultraviolet radiation of gas-shielded welding operations. This requirement is to avoid the decomposition of solvent vapors into toxic substances.

FIRE/EXPLOSION VENTILATION STANDARDS

Federal OSHA standards contain requirements for ventilation aimed at preventing formation of flammable or explosive gas or vapor concentrations. The standards cover both specific operations and specific materials. The fire and explosion prevention ventilation standards for welding, cutting, and brazing were discussed earlier in this chapter, along with the health ventilation standards for these operations.

Spray Finishing

Spray finishing is defined as "methods wherein organic or inorganic materials are utilized in dispersed form for deposit on surfaces to be coated, treated or cleaned." Standards appear in Parts 1910.94c and 1910.107.

For any spray-finishing operation the lights, motors, electrical equipment, and other sources of ignition, including parts of the ventilation system, must meet OSHA standards for explosive atmospheres in Part 1910.107c. The construction of the ventilation system must recognize that sprayed materials will accumulate on ducts, fans, and other components if not removed by filters. An accumulation would add to the weight of ducts and could impair fan performance. Duct inspection and cleaning doors are important design considerations.

All spraying areas must be provided with mechanical ventilation to safely remove flammable vapors, mists, or powders and to control combustible residues so that no hazards result. Each spray booth requires an independent exhaust duct discharging outside the building. However, multiple booths using identical spray materials may have a common exhaust if their combined open frontal area does not exceed 18 ft.² If more than one fan serves a booth, the fans must be interlocked so that they all run together. Ductwork used to ventilate spray-finishing operations cannot be connected to ducts that exhaust other processes, chimneys, or flues.

To prevent sparks in the ventilation system, the fan-rotating element must be nonferrous or nonsparking; or the fan case can be made of or be lined with nonsparking material. Since fan parts expand and sprayed materials accumulate, there must be enough clearance between the fan blade and housing to avoid fire from friction. The fan shaft and bearings must be heavy enough to maintain alignment even when the blades are loaded. Self-lubricating bearings or bearings that are lubricated from outside the duct are needed. Exhaust fan motors must not be placed inside the spray booths or ducts. If fan belts enter the booths or ducts, they must be enclosed within the booth or duct.

Although exhaust ducts without dampers are preferred, ducts with dampers must be fully open when the ventilation system is in operation. Exhaust ducts require a clearance from unprotected combustible construction or other combustible material of not less than 18 inches. (This distance can be reduced if automatic sprinklers or fire protection construction is provided.)

Unless the spray booth exhaust duct is from a water wash, the discharge must not be within 6 ft of any combustible exterior walls or roof nor in the direction of any combustible construction or unprotected opening within 25 ft. Exhaust discharges must not contaminate make-up air and cannot be recirculated. Adequate room intake openings located to minimize dead air pockets should be used.

Regardless of the spray-finishing method (air-operated guns, airless or electrostatic sprayers), sufficient ventilation is needed to prevent accumulation of flammable vapors. Except where a spray booth has an adequate separate supply of replacement air, the face velocity into all openings must meet the criteria presented in Figure 3.14. An adequate air replacement system for a spray booth is one that introduces air directly into the booth above or upstream of the object being sprayed so that the minimum velocity past the object meets the values in Figure 3.14.

In addition, the exhaust rates must be sufficient to dilute the solvent vapor from spraying to no more than 25% of the lower explosive limit (LEL). To calculate the dilution air volume, use this equation from the OSHA standard:

Operating Conditions for Objects Completely Inside Booth	Crossdraft fpm	Airflow Velocities, fpm	
		Design	Range
Electrostatic and automatic airless operation contained in booth without operator	Negligible	50 large booth	50-75
		100 small booth	75-125
Air-operated guns, manual or automatic	Up to 50	100 large booth	75-125
		150 small booth	125-175
Air-operated guns, manual or automatic	Up to 100	150 large booth	125-175
		200 small booth	150-250

Notes:
 (1) Attention is invited to the fact that the effectiveness of the spray booth is dependent upon the relationship of the depth of the booth to its height and width.
 (2) Cross drafts can be eliminated through proper design and such design should be sought. Crossdrafts in excess of 100 fpm should not be permitted.
 (3) Excessive air pressures result in loss of both efficiency and material waste in addition to creating a backlash that may carry overspray and fumes into adjacent work areas.
 (4) Booths should be designed with velocities shown in the column headed "Design." However, booths operating with velocities shown in the column headed "Range" are in compliance with this standard.

Figure 3.14 Federal OSHA spray booth air velocity standards.

$$\text{Dilution Air, ft}^3/\text{gal} = \frac{4\,(100\text{-LEL}) \times \text{Vapor Released, ft}^3/\text{gal}}{\text{LEL}} \quad (3.3)$$

where LEL = lower explosive limit, percent

To find dilution airflow rate (ft³/min), multiply the dilution air from Equation 3.3 by the gallons of solvent used per minute.

Table 3.5 lists the LEL and vapor volume released for some common solvents. For solvents not listed in Table 3.5, the vapor volume released can be calculated from the molecular weight and specific gravity:

$$\text{Vapor Released, ft}^3/\text{gal} = \frac{3224 \times \text{sp gr}}{M} \quad (3.4)$$

where sp gr = specific gravity of liquid solvent (water = 1.0)
      M = molecular weight of solvent

Example: A lacquer spraying operation uses 3 gallons of methyl acetate solvent per hour. Calculate the dilution air requirement.

Answer: From Table 3.5 the lower explosive limit for methyl acetate is 3.1% and 40 ft³ of vapor are released per gallon. Using Equation 3.3:

$$\text{Dilution air, ft}^3/\text{gal} = \frac{4\,(100 - 3.1) \times 40\ \text{ft}^3/\text{gal}}{3.1}$$

$$= 5001.3\ \text{ft}^3/\text{gal}$$

Table 3.5   Vapor Volume and LEL of Solvents

Solvent	Vapor Volume, ft³ of vapor/gallon of liquid at 70°F	Lower Explosive Limit, percent by volume at 70°F
Acetone	44.0	2.6
Amyl Acetate (*iso*)	21.6	1.0[a]
Amyl Alcohol (*n*)	29.6	1.2
Amyl Alcohol (*iso*)	29.6	1.2
Benzene	36.8	1.4[a]
Butyl Acetate (*n*)	24.8	1.7
Butyl Alcohol (*n*)	35.2	1.4
Butyl Cellosolve	24.8	1.1
Cellosolve	33.6	1.8
Cellosolve Acetate	23.2	1.7
Cyclohexanone	31.2	1.1[a]
1,1 Dichloroethylene	42.4	5.9
1,2 Dichloroethylene	42.4	9.7
Ethyl Acetate	32.8	2.5
Ethyl Alcohol	55.2	4.3
Ethyl Lactate	28.0	1.5[a]
Methyl Acetate	40.0	3.1
Methyl Alcohol	80.8	7.3
Methyl Cellosolve	40.8	2.5
Methyl Ethyl Ketone	36.0	1.8
Methyl n-Propyl Ketone	30.4	1.5
Naphtha (VM&P) (76° Naphtha)	22.4	0.9
Naphtha (100° F Flash) Safety Solvent-Stoddard Solvent	23.2	1.0
Propyl Acetate (*n*)	27.2	2.8
Propyl Acetate (*iso*)	28.0	1.1
Propyl Alcohol (*n*)	44.8	2.1
Propyl Alcohol (*iso*)	44.0	2.0
Toluene	30.4	1.4
Turpentine	20.8	0.8
Xylene (*o*)	26.4	1.0

[a]at 212°F.

Since 3 gallons are used per hour the airflow requirement is

$$5001.3 \ \frac{\text{ft}^3}{\text{gal}} \times \frac{3 \ \text{gal}}{\text{hr}} \times \frac{\text{hr}}{60 \ \text{min}} = 250 \ \text{ft}^3/\text{min}$$

A nomograph (Figure 3.15) is available to aid in solving Equations 3.3 and 3.4.[12] If the vapor volume released and LEL of the solvent are known, connect the proper values in columns 5 and 3 and extend the line to column 2 to read the answer. If only the solvent specific gravity, molecular weight and LEL are known, connect the values in columns 1 and 4 and continue the straight line to find the vapor volume released in column 5. Then connect the vapor volume released with the LEL in

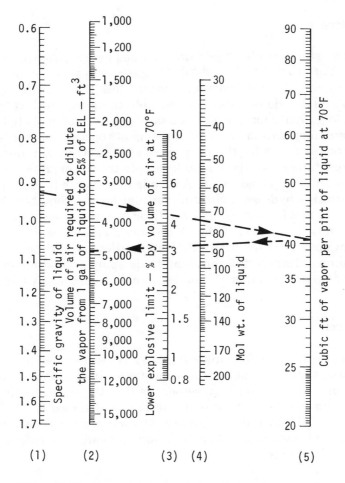

Figure 3.15 Nomograph for determining spray booth airflow to meet federal OSHA standards. (Source: Reference 12)

column 3 and extend the straight line to find the dilution air in column 2.

Example: Assume that for the previous example the vapor released per gallon is not known. The molecular weight for methyl acetate is 74.08 and the liquid specific gravity is 0.927.[11] Use the nomograph in Figure 3.15 to find the dilution rate.

Answer: As illustrated in Figure 3.15 connect the specific gravity in column (1) with the molecular weight in column (4) and extend the line to column (5) to find the volume of vapor released. Then connect the value in column (5) with the LEL in column (3) and extend the line to column (2). The answer: 4900 ft³ of air is needed to dilute the vapors from one gallon of methyl acetate to 25% of the LEL.

Hydrogen

Ventilation standards for liquid and gaseous hydrogen systems are contained in Part 1910.103. Standards for gaseous systems do not apply to systems having a total hydrogen content less than 400 ft.^{3} Liquefied system standards do not apply to portable containers of less than 150-liter (39.63-gallon) capacity. The standards also do not apply to hydrogen manufacturing plants or other establishments operated by the hydrogen supplier or his agent for the purpose of storing hydrogen and refilling portable containers, trailers, mobile supply trucks, or tank cars.

Generally the standards require adequate vents to permit safe release of leaking hydrogen. Natural ventilation is acceptable. Air inlets to the room must be located near the floor in exterior walls only. Outlet openings must be located at the high point of exterior walls or the roof. Minimum open area for both inlets and outlets is 1 ft^2 per 1000 ft^3 of room volume. Slightly different standards apply depending on the size of the system and whether it handles liquid or gaseous hydrogen.

Oxygen

Ventilation standards for bulk oxygen systems installed on industrial and institutional consumer premises are contained in Part 1910.104.

The only ventilation requirement is that the storage system must be located to insure adequate ventilation. A distance of 75 ft in one direction and 35 ft in approximately a 90° direction from confining walls not including firewalls less than 20 ft high is required in courtyards and similar confining areas.

Flammable and Combustible Liquids

Standards for flammable and combustible liquids appear in Part 1910.106 of the federal OSHA standards.

For inside storage of these liquids in drums or other containers not exceeding 60-gallon individual capacity or in portable tanks not exceeding 660-gallon capacity, a gravity or mechanical ventilation system is required (Part 1910.106d). The system must provide at least six air changes per hour. If a mechanical system is used, it must be controlled by a switch located outside of the door, and the system and any lights must be operated by the same switch. A pilot light adjacent to the switch is needed if Class I flammable liquids (flash point below 100°F) are dispensed. For gravity ventilation the fresh air intake and exhaust outlet must be on the exterior of the building in which the room is located.

Industrial plants where the use of flammable or combustible liquids is incidental to the principal business, or where they are involved only in unit operations such as mixing, drying, or similar operations not involving chemical reaction, must comply with standards in Part 1910.106e. For incidental use, adequate natural or mechanical ventilation is required.

For unit operations involving Class I liquids a ventilation rate of at least 1 ft³/min per ft² of solid floor area is required. Either natural or mechanical ventilation can be used with the exhaust discharged at a safe location outdoors. Make-up air, introduced to prevent short-circuiting the exhaust, is needed. Ventilation must include all floor areas or pits where flammable vapors may collect. The system must limit flammable vapor/air mixtures to the interior of the equipment. In the case of air exposure of Class I liquids, the flammable mixture must be confined within a 5-ft radius of the equipment. The same standard also applies to process plants handling flammable or combustible liquids (Part 1910.106h).

Bulk plants handling Class I liquids require ventilation in all rooms, buildings, or enclosures where the liquids are pumped or dispensed (Part 1910.106f). Where natural ventilation is inadequate, mechanical ventilation must be provided. If Class I liquids are handled inside a building with a basement or pit allowing accumulation of flammable vapors, ventilation is needed. Filling or drawing from containers of Class I liquids is allowed only if provision is made to prevent a hazardous accumulation of vapors. If mechanical ventilation is used, it must be kept in operation while flammable liquids are being handled.

Service stations may not store or handle Class I liquids inside a building having a basement or pit where flammable vapors may travel unless these areas are provided with ventilation. Inside service stations (dispensing areas located within buildings) require a mechanical or natural ventilation system if above or at grade. Below grade, a mechanical system and automatic sprinklers are required. The fan must be electrically interlocked with the gasoline pump to prevent its operating if the fan is turned off.

SUMMARY

This chapter outlined the federal OSHA standards requiring ventilation. These standards may change in the future. In addition, some states have their own occupational safety and health standards. The following chapters will help you to determine which areas or processes require ventilation, other methods of reducing employee exposures, and the best way to provide the ventilation.

REFERENCES

1. "OSHA General Industry Safety and Health Regulations," U.S. Code of Federal Regulations, Title 29, Chapter XVII, Part 1910.
2. U.S. Code of Federal Regulations, Part 1910.252.
3. State of Oregon Occupational Health Regulations, Appendix B, Section 22–1017(A).
4. Rosensted, R.E., S.K. Shama, and J.P. Flesch. "Occupational Health Case Report Number 1," *Journal of Occupational Medicine* 16, No. 1, 22.
5. American National Standard Z 9.2. "Fundamentals Governing the Design and Operation of Local Exhaust Systems" (New York, New York: American National Standards Institute, 1960).
6. American National Standard Z 33.1. "Installation of Blower and Exhaust Systems for Dust, Stock, and Vapor Removal or Conveying" (New York, New York: American National Standards Institute, 1961).
7. American National Standard Z 43.1. "Ventilation Control of Grinding, Polishing and Buffing Operations" (New York, New York: American National Standards Institute, 1966).
8. American National Standard B 7.1. "Safety Code for the Use, Care and Protection of Abrasive Wheels" (New York, New York: American National Standards Institute, 1970).
9. ACGIH Committee on Industrial Ventilation. *Industrial Ventilation—A Manual of Recommended Practice,* 17th Ed. (Lansing, Michigan: American Conference of Governmental Industrial Hygienists, 1982).
10. Bastress, E.K., J.M. Niedzwecki, and A.E. Nugent. "Ventilation Requirements for Grinding, Buffing and Polishing Operations," U.S. Department of Health, Education and Welfare, Publication No. (NIOSH) 75–107.
11. Sax, N.I. *Dangerous Properties of Industrial Materials* (New York, New York: Van Nostrand Reinhold Company, 1975).
12. Sisson, W. "OSHA Spray Booth Air Flow Requirements," *Pollution Engineering* 7, No. 11, 31 (1975).

Note: For References 5, 6, 7 and 8 the version cited in Federal OSHA standards is listed; for current design and operating information consult the latest edition.

CHAPTER 4

Hazard Assessment

As discussed in the previous chapter, OSHA standards require ventilation for certain specific operations such as arc welding in small workrooms or when opening closed systems in which certain carcinogens are handled. For operations not specifically mentioned in OSHA standards, ventilation or another control method is needed only if exposures to workers exceed the maximum OSHA exposure limits for the substances, or if there are indications of harmful exposures whether or not a legal exposure standard exists. In order to avoid the need for medical monitoring of employees or periodic air sampling, ventilation is also useful in further reducing some exposures that do not exceed OSHA allowable limits. The decision whether or not ventilation is needed when it is not specifically required by OSHA standards should be based on a study that identifies the exposures that occur and compares them to toxicological standards.

Accurate determination of exposures and toxicological evaluations are part of industrial hygiene. This chapter is designed to help you use good industrial hygiene techniques to determine whether a hazardous exposure exists. Proper determination and documentation of nonhazardous exposures allow you to concentrate your efforts on the exposures that can cause problems. A hazard assessment usually follows these steps (Table 4.1):

Table 4.1   The Hazard Assessment Process

- Identify the chemicals and contaminants.
- Find the allowable exposure to each substance.
- Determine source and time duration of exposures.
- Determine airborne concentration of each contaminant.
- Calculate average and peak exposures.
- Compare employee's exposure to allowable exposure standard.

- Identify the chemicals and contaminants in the process.
- For each contaminant find the allowable OSHA exposure standard or other safe exposure guidelines based on the toxicological effect of the material.
- Understand the industrial process well enough to see where contaminants are released, and where and how long employees are exposed.
- Use air sampling techniques to determine the level of airborne contaminants.
- Calculate the resulting daily average and peak exposures from the air sampling results and employee exposure times.
- Compare the calculated exposures with OSHA standards and the Threshold Limit Values (TLV) listing published by the American Conference of Governmental Industrial Hygienists [1], Hygienic Guides [2] or other toxicological recommendations.

The level of difficulty of the hazard assessment process depends on the complexity of the industrial operation, the number of different contaminants involved, and the ease of analyzing the workroom air for each of the contaminants. Industrial hygiene assistance is available from insurance carriers, government occupational health consultant departments or similar nonenforcement agencies, and private consultants. Occasionally the company supplying your raw materials can also provide some assistance. It may be advisable to have a detailed survey by an experienced industrial hygienist before making large expenditures to install or upgrade ventilation systems. But safety and health personnel in most plants should be equipped and trained to perform hazard assessments for the routine processes in their plant.

IDENTIFYING THE CONTAMINANT

There are four major sources of contaminants in an industrial operation (Table 4.2):

- Chemicals or other substances that are the raw materials for the process or part of the operation itself. For example, paint thinner is a raw material in paint manufacturing. Likewise, a solvent vapor degreaser will emit vapors of the solvent used in the degreaser, perhaps 1,1,1-trichloroethane (also called methyl chloroform). Possible impurities in raw materials such as phosphine in the some carbide-generated acetylene gas [3] should also be considered.
- Intermediate and final products formed from the raw materials. This is an obvious source of contaminants whenever a chemical reaction

Table 4.2   Sources of Contaminants

Source	Example
Raw materials or process chemicals	Electroplating, solvent cleaning, silica from sandblasting
Intermediates or final products	Pesticides, compressed gases
Undesirable chemical reactions	Carbon monoxide from incomplete combustion, ozone from sparking electrical contacts
Inadvertent breakdown of chemicals	Phosgene from arc welding around chlorinated solvents

occurs, but the identity of intermediates and their toxicological properties may be difficult to determine.

• Undesirable chemical reaction products formed during the operation. Since they are formed by chemical reactions, they are different from the raw materials. The most common contaminant in this category is carbon monoxide (CO) from incomplete combustion of organic material. Thus, any process involving combustion is a potential source of CO contamination if the combustion products are not exhausted outside of the work area. Another example of an undesirable chemical reaction is the zinc oxide fumes formed when galvanized (zinc-coated) steel is heated during welding or torch cutting. The high temperatures cause the zinc to combine with oxygen to produce zinc oxide. Breathing too much zinc oxide causes a well-known flu-like disease with chills, cough, and general malaise that goes away after about one day with no recognized lasting effects. The condition is not caused by finely divided zinc metal, only zinc oxide.

• Inadvertent breakdown of chemical compounds into more toxic or more irritating substances. Probably the best example of this category is the danger of generating phosgene gas by welding near a vapor degreaser containing a chlorinated hydrocarbon solvent such as trichloroethane or perchloroethylene. The ultraviolet light from the welding arc breaks the solvent down into phosgene, hydrochloric acid, and other compounds. Phosgene, a severe lung irritant, is one of the most acutely toxic gases known and is harmful at very low concentrations. Another example is accidental release of hydrogen cyanide gas if enough acid is mixed with cyanide electroplating salts or solutions to lower the pH to the acidic range (less than pH 7.0). This is a safety consideration whenever plating tank leakage, earthquakes, or other circumstances could result in acid mixing with cyanide plating solutions.

Regardless of source of contamination, the first step is to identify each contaminant. When pure chemicals or widely known compounds are used, identification may be easy. When proprietary mixtures of chemicals are encountered, such as tradename products, identification may be difficult unless the manufacturer is willing to supply information on the product. If you expect to collect air samples to determine exposure levels, you will probably need the exact names of chemicals in the product to determine proper sampling and analytical procedures. The best source of information on product composition is the Material Safety Data Sheet (MSDS). Another source is *Clinical Toxicology of Commercial Products* by Gleason et al.[4] Although intended primarily for treatment of accidental oral poisoning, this text contains valuable information for identifying possible airborne contaminants from commercial products. Laboratory analyses of bulk product samples to determine the individual components is the last resort. These analyses are expensive, especially for complex mixtures.

ALLOWABLE EXPOSURE LIMITS

Once the identity of the chemical contaminants is known, the allowable exposure limit for each is determined. This is done before air sampling or other exposure measurements are performed because in some cases the air monitoring technique depends on the way the allowable limit is defined. For example, the current TLV for oil mist is 5 mg/m^3 with a footnote that reads "as sampled by a method that does not collect vapor."[1] Thus for the monitoring result to be validly compared to this TLV, care must be taken to avoid collecting and analyzing any oil vapor that is present along with the oil mist (droplets).

Often there will not be established exposure criteria for all contaminants. Although there are adopted TLVs for over 600 substances or chemical categories (nickel compounds, vanadium compounds, etc.), there are many times that number of specific chemicals in use within industry. If there are no TLVs or OSHA standards for the specific substances, consult manufacturer's Material Safety Data Sheets or other product literature for exposure recommendations. It may be helpful to refer to toxicology references [5] or contact the supplier directly to see whether they have set internal exposure criteria for their own operations.

When no exposure criteria can be found or estimated for a chemical substance, there is often no immediate value in measuring its airborne levels as part of a hazard assessment.

SOURCE AND DURATION OF EXPOSURE

Since the workers' overall or average exposure to a contaminant depends on both the concentration and length of exposure, you need to know enough details about the industrial process to develop this information. Often the plant engineering or plant safety personnel have an advantage over outside consultants in this phase of the hazard assessment. The plant staff can be familiar with the different plant jobs, upset conditions, and other factors that affect exposures.

The goal is to identify the contaminants generated during each step of the process and to describe the location and work times of personnel that might be exposed to the contaminants. The easiest jobs to define are assembly line or similar repetitive operations where a worker remains in one location and repeats the same tasks over and over during the day. The process description involves measuring the time schedule of each repetition with emphasis on when different contaminants are formed or escape into the workroom, and then counting the repetitions per day. This procedure is used for each different position on the assembly line if needed as part of the hazard assessment for the entire plant. Of course, few jobs are this easy to describe. In most plants the specific tasks vary from day to day depending on work assignment, the products being made, machinery availability, and other factors.

When process variations make it difficult to define typical exposures, two approaches are used:

- Separate descriptions can be prepared for all or most of the exposure patterns that are encountered. The number can vary from a few to too many; the total number often determines whether this approach is feasible.
- A "worst case" exposure pattern can be selected to represent the longest exposure times to the highest expected contaminant levels. Even if this exposure pattern occurs only infrequently, it provides a convenient starting point for evaluating employee exposures. If air sampling or other evaluation shows that the "worst case" is not hazardous, then there is no problem with the other exposures. If the "worst case" does exceed allowable average or peak exposure criteria, then other lesser exposures can be defined and evaluated until an acceptable case is found. Then controls can be developed to reduce the excessive exposures.

Example: In an electronic printed circuit (PC) board fabrication shop, workers prepare PC boards by coating copper-covered resin sheets with wax over the circuit layout. When the board is etched in acid, the wax protects the circuit while the excess copper is removed. Finally the wax is removed in a solvent spray cleaner containing 1,1,1-trichloroethane. There is an odor of solvent around the spray cleaner and in the general work area where the boards are prepared.

Usually three employees share the work equally, but occasionally one employee may clean all 25 PC boards produced per day. Each board is cleaned for 10 minutes in the solvent cleaner. The employees spend their time in the general work area when not at the solvent cleaner. What is the "worst case" exposure pattern?

Answer: The "worst case" occurs when a single employee cleans all 25 PC boards in the solvent cleaner:

$$\frac{25 \text{ boards}}{\text{day}} \times \frac{10 \text{ min}}{\text{board}} = 250 \text{ min/day at the cleaner}$$

For an 8-hr shift

$$\frac{480 \text{ min}}{\text{shift}} - 250 \text{ min} = 230 \text{ min/day in the general work area}$$

With this breakdown the "worst case" employee exposure can be estimated by the method illustrated later in this chapter (page 93).

MEASURING AIRBORNE CONTAMINANTS

After you have identified the contaminants and know when and where employees may be exposed to them, you are ready to determine how much of each contaminant is in the air. Sometimes a chemical has a distinctive odor well below any level with health implications. If you cannot smell the contaminant there is probably no health hazard. Ammonia is an example: most people can smell ammonia at 5–10 ppm, [5] which is well below current exposure limits. If you cannot smell ammonia there is probably no problem. If you can smell ammonia you do not know whether the concentration is above or below acceptable limits unless air samples are taken.

For some other exposures you can rule out a potential health hazard based on the amounts of contaminants used or generated in the process. For example, occasional spray painting in a large, well-ventilated room with small aerosol spray cans probably does not present a health hazard. The same is not necessarily true in a smaller room without adequate general ventilation. The dilution ventilation calculation formula in Chapter 1 can help with this type of evaluation.

Except for the cases mentioned above, some air sampling is usually needed to determine contaminant levels. Air sampling can be either the hardest or the easiest part of the hazard assessment. Simple and inexpensive measuring techniques suitable for field use are available for many toxicants; for other materials air samples must be collected and returned to a laboratory for analysis. Each plant should have the capability to sample for contaminants that may be present at levels high enough to represent significant exposures to employees.

Sampling Guidelines

This chapter is not intended as a detailed review of industrial hygiene sampling techniques. However, here are some guidelines to help estimate

Table 4.3 Air Sampling Guidelines

- Collect breathing zone samples.
- Collect "peak exposure" samples to see whether highest concentration is below acceptable limits.
- Calibrate or test the sampling equipment.

the level of airborne contaminants to determine if controls are needed or if existing controls are working properly (Table 4.3):

• Plan and carry out the air sampling to reflect the actual exposure that the workers receive. The sampling should cover as much of the work shift or exposure period as possible. Usually "breathing zone" samples should be collected (Figure 4.1) to determine the contaminant level at the worker's face. General room or area samples are helpful in assessing whether employees working nearby are exposed even though they are not directly involved in the operation.

Breathing zone samples are needed because a worker's actual exposure is reduced every time he moves away from the process. Conversely, the worker's exposure is higher when he bends over the source. Air samples that do not take the worker's movements into account do not accurately reflect his exposure. The best way to determine the capture efficiency of a ventilation hood is to collect stationary samples near the hood when the ventilation is on and when it is off. But using the stationary test results to extrapolate whether or not employee exposures exceed OSHA limits can be risky.

Figure 4.1 Continuous breathing zone air samples are most accurate for determining workers' exposure to airborne contaminants. This sample is being collected on a tube of activated carbon (clipped to collar) using a battery-powered air pump.

- Collect "peak exposure" samples first to rule out no-problem exposures. In many operations there are peak exposure periods, such as when loading dusty raw materials into the process or opening an enclosure to remove the final product. If the contaminant levels during these peak periods are too low to be a problem (for example, if the peak is lower than the allowable 8 hr average exposure standard), then the lower levels throughout the rest of the operation are not a problem either.
- Calibrate or test the sampling equipment. Regardless of the sampling equipment you use, it must be tested or calibrated periodically so that you have confidence in the results. If you are using a hand pump and indicator tubes (Figure 4.2), test the pump for leakage by following the manufacturer's instructions. Calibrate electrically powered pumps to determine the airflow through the filter or charcoal adsorption tube used to collect the sample (Figure 4.3). Use a known mixture of gas or vapor (Figure 4.4) to test direct-reading instruments measuring gas or vapor concentration or explosive vapor ranges. Checking for explosive mixtures with an uncalibrated instrument can be hazardous to your health. Internal (or built-in) flow indicators and calibration checks are no substitute for independent tests of the sampling device's accuracy. Calibration and testing techniques are available for most industrial hygiene instruments.
- Record enough data about contaminants, sampling equipment, location, date, time, person sampled, work activities, and unusual occurrences to reconstruct the events accounting for the exposure. A data sheet (Figure 4.5) helps you to avoid overlooking needed information when in the field.

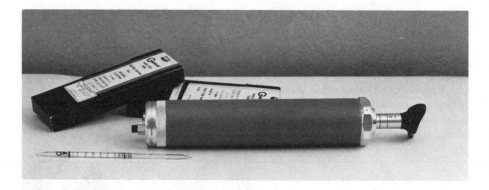

Figure 4.2   Indicator, or detector tubes, and hand pump make up the most widely used direct-reading sampling device. Tubes, which indicate concentration by color change or length-of-stain, are available for many common contaminants.

Figure 4.3 A burette-bubble meter is a convenient way to calibrate pumps. The distance that a soap bubble moves in a measured time period indicates the pump flow rate.

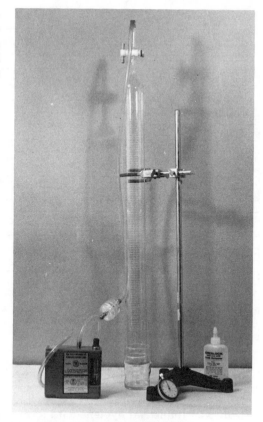

Figure 4.4 Known concentration gas sources are needed to calibrate gas and vapor detectors. This is a carbon monoxide detector with calibration gas.

INDUSTRIAL HYGIENE SAMPLING FORM		PAGE 1 OF 1	SAMPLED BY GRF	SAMPLE NO. 200
EMPLOYEE (NAME & NO.) Jack Johns	PLANT/FACILITY Printed circuit fabrication		ACCOUNT NO. 123-47	DATE 7/4/76
JOB CLASSIFICATION Technician	CONTAMINANT 1,1,1-Trichloroethane		TYPE SAMPLE ▶	[X] PERSONAL [ ] AREA [ ] SOURCE

AREA SAMPLED
Fabrication Room including solvent spray cleaner and general work area

SKETCH AREA: SHOW TEST LOCATIONS, PERSONNEL, SOURCES, WIND DIRECTIONS, VENTILATION, NORTH DIRECTION.

WEATHER CONDITIONS	WIND- NA		TEMPERATURE- NA
	BAROMETER- NA		HUMIDITY- NA
SOURCE DISTANCE	TO GROUND NA	TO EMPLOYEE NA	

TIME STARTED	WORK TASKS	TIME AT TASK
8:00-9:55A	Sample 1 - Cleaned 4 boards	1hr,55min
9:55-12:05P	Sample 2 - Cleaned 6 boards	2hr,10min
12:05-2:00P	Sample 3 - Cleaned 3 boards	1hr,55min
2:00-4:00P	Sample 4 - Cleaned 7 boards	2hr
2:13	Peak sample - duplicate sampling to Sample 4	12min
	while cleaning one board	

INSTRUMENT TYPE & NO. Sipin·SP-2		FLOW RATE 40 cc/min	VOLUME SAMPLED --
CALIBRATED BY GRF	DATE 7/4/76 METER READING INITIAL: -- FINAL: --		DETECTOR TUBE READING --
REMARKS: Strong odor of solvent around cleaner unit.			

Sketch: North. Solvent cleaner. Work bench, Work bench, Cabinets, Windows closed, Door open, Hallway.

LAB ANALYSIS	DATE	BY
See below PPM mg/m³	7/19/76	Peters
REMARKS: Sample 1 - 127 ppm Peak = 243 ppm		
Sample 2 - 153 ppm		
Sample 3 - 98 ppm .		
Sample 4 - 175 ppm		

Figure 4.5 Industrial hygiene data form helps you to record the necessary information about each sample.

Sampling Techniques

Air sampling techniques fall into two main categories: direct-reading instruments that give you results immediately and sampling devices that collect a sample in the field for later laboratory analysis. Although direct-reading instruments are handy, they often collect only short-term, or "grab," samples. Samples collected for laboratory analysis using low flow rate pumps cover longer time periods. Once this distinction is made then the type of contaminant becomes important. Characteristics of air contaminants are discussed in Chapter 6, but your choice for a sampling technique depends on whether contaminants are particulates, gases, or vapors (Table 4.4):

- Particulate matter, including dust, fumes, mists, and smoke, is usually collected on filters or in liquid-filled impingers that retain the

Table 4.4 Air Sampling Techniques

For Particulates
  Collection on filter media for laboratory weighing or chemical
    analysis.
  Collection in impingers containing liquids that trap particulates for
    laboratory counting or chemical analysis.
  Direct-reading instruments for respirable and total dust.

For Gases and Vapors
  Indicator tubes.
  Direct-reading instruments.
  Adsorption on activated carbon or other material for laboratory
    analysis.
  Absorption in a liquid media contained in a bubbler for laboratory
    analysis.

particulates for later analysis. Small battery-powered pumps are often used for personal sampling. Although some direct-reading instruments are available for total and respirable dust, chemical analysis may still be needed to determine how much of each contaminant is in the air. Rely on the analytic laboratory for recommendations on collection devices (type of filter or collecting solution), sampling rate and time, and other parameters.

• Gases and vapors are measured either by direct reading instruments or by collection for later analysis. Indicator tubes (Figure 4.2) are the most popular direct-reading device and are available for a variety of different contaminants from different manufacturers. One rule to observe, however, is not to mix different brand tubes and sampling pumps. Although this may seem harmless, tests showed large errors resulted from different airflow rate and volume between pumps.[6] In addition to indicator tubes a wide selection of instruments working on heat of combustion, infrared absorption, ionization, and other principles are available.[7]

For sampling many organic vapors, adsorption on activated carbon is increasingly popular. The sample is collected with a low flow rate pump and a tube containing the carbon (Figure 4.1) or with a badge-like device called a passive dosimeter. The dosimeter collects the vapors by diffusion without a pump.

At the laboratory the chemicals trapped on the carbon are removed and analyzed. However, since not all gases and vapors are retained on the carbon, specific guidelines should be sought from the analytical laboratory. Gas and vapor samples can also be collected for analysis in special bags or collecting solutions.

Sampling Time

The best samples are long-term continuous samples covering the entire exposure period or work shift. Short-term samples during peak exposure periods help quantify the highest exposures to determine whether they exceed OSHA standards or present a health hazard. When you can sample for all of the work shift, you eliminate the major unknown of what happened during the time when you were not sampling. With indicator tubes and other grab sampling techniques it is not practical to take enough samples to account for the entire work shift. In these cases you cannot avoid assuming that the samples represent either the typical or peak exposure (depending on how the sample was collected) for the time period. Random sampling throughout the day helps avoid sampling bias for grab samples.

Example: Design a sampling plan to determine employee exposure to 1,1,1-trichloroethane in the printed circuit board shop described earlier in this chapter.

Answer: Perform sampling over as much of the 8-hour workday as possible, plus a short-term sample during actual operation of the spray cleaner to estimate the peak exposure level. Record the number of boards that are cleaned during each sampling period so that those results can be used to estimate exposures on days when the workload may be heavier.

After consultation with the analytical laboratory, four separate samples covering about 2 hrs each were collected to minimize the risk of overloading the activated carbon sampling device. A 12-minute short-term sample was also collected. The following data were collected:

Sample No.	Time Period	Elapsed Time, hr:min	Boards Cleaned	1,1,1-Trichloroethane Level, ppm
1	8:00 A.M.– 9:55 A.M.	1:55	4	127
2	9:55 A.M.–12:05 P.M.	2:10	6	153
3	12:05 P.M.– 2:00 P.M.	1:55	3	98
4	2:00 P.M.– 4:00 P.M.	2:00	7	175
			20	
Peak	2:13 P.M.– 2:25 P.M.	0:12	1	243

These data are recorded on the industrial hygiene data sheet (Figure 4.5).

CALCULATING EXPOSURES

After the air sampling results are known, you can calculate the employee's Time Weighted Average (TWA) exposure for the day. The significance of the TWA exposure was discussed in the previous chapter and is the exposure standard in OSHA regulations for most contaminants. The TWA is calculated using this equation:

$$TWA = \frac{C_1T_1 + C_2T_2 + \ldots + C_nT_n}{8 \text{ hr}} \qquad (4.1)$$

where TWA = time weighted average concentration, usually ppm or mg/m^3.

C = concentration of contaminant during the exposure period, same units as TWA.

T = time duration of corresponding exposure period, hr.

Note that 8 hr is used as the denominator since OSHA standards and TLVs are based on an 8-hr work day.

If the contaminant has a ceiling or peak exposure standard, then select the sample results corresponding to the employee's greatest short-term exposure to determine whether this peak exposure exceeds allowable limits.

Example: Calculate the TWA exposure for the data in Figure 4.5. Also recall that a "worst case" exposure was defined earlier for a day when a single employee cleaned 25 printed circuit boards. Estimate the employee's "worst case" exposure.

Answer: Using Equation 4.1:

$$TWA = \frac{(127 \text{ ppm})(1.9 \text{ hr}) + (153 \text{ ppm})(2.2 \text{ hr}) + (98 \text{ ppm})(1.9 \text{ hr}) + (175 \text{ ppm})(2 \text{ hr})}{8 \text{ hr}}$$

$$TWA = \frac{241.3 + 336.6 + 186.2 + 350.}{8} = 139 \text{ ppm}$$

Also, from Figure 4.5 the peak exposure was 243 ppm.

On this day the worker cleaned 20 boards. On the "worst case" day he could clean 25 boards. An estimate of his exposure is:

$$\frac{25 \text{ boards}}{20 \text{ boards}} \times 139 \text{ ppm} = 174 \text{ ppm}$$

If a few grab samples using indicator tubes were collected rather than continuous samples, you would estimate the "worst case" exposure using the work pattern information on Page 86. Suppose 10 samples collected during the day showed the average concentration at the solvent cleaner was 225 ppm while the average concentration in the general work area was 55 ppm. The "worst case" is:

$$TWA \approx \frac{(250 \text{ min/shift at cleaner}) \times 225 \text{ ppm} + (230 \text{ min/shift in workroom}) \times 55 \text{ ppm}}{480 \text{ min/shift}}$$

$$TWA \approx \frac{56,250 + 12,650}{480} \approx 144 \text{ ppm}$$

COMPARE EXPOSURE TO
ALLOWABLE LIMITS

To determine whether an overexposure exists, compare the measured exposure levels to the ACGIH Threshold Limit Values or the OSHA exposure standards. The federal OSHA standards for air contaminants are listed in Title 29, Chapter XVII, Part 1910.1000 of the *Code of Federal Regulations*. State OSHA plans have similar standards in their state administrative codes. As discussed in the last chapter, only an 8-hr TWA OSHA standard exists for some contaminants; for others, ceiling concentration standards and short-term peaks above the ceiling have also been specified. Some state standards are more specific than the federal standard in establishing peak exposure limits.[8] For some materials, exposure below the legal standard may also require medical surveillance, periodic air sampling, and other steps. To avoid the monitoring and other requirements, it may be advantageous to use ventilation or another control to reduce exposures below these levels.

Example: The federal OSHA standard for exposure to 1,1,1-trichloroethane (also called methyl chloroform) is 350 ppm.[9] Do exposures in the printed circuit shop exceed these levels?

Answer: The calculated TWA was 139 ppm on the sampling day and the estimated "worst case" exposure was 144 or 174 ppm, depending on calculation method. Since all of these data are below the 350 ppm OSHA limit, no violation exists in this shop based on that day's exposures.

TLVs and OSHA exposure standards do not exist for all contaminants. The goal is to protect employees' health, not just meet a legal standard. Regardless of established limits, evidence of adverse health effects from occupational exposures means that controls are needed to reduce exposures.

SUMMARY

The hazard assessment steps outlined in this chapter provide a guide for assessing whether exposure problems exist in a plant and what their possible sources might be. Once you know which exposures pose potential health hazards or exceed legal standards you can concentrate on reducing those exposures to acceptable levels.

REFERENCES

1. American Conference of Governmental Industrial Hygienists. "Threshold Limit Values for Chemical Substances in Workroom Air" (Cincinnati, Ohio: ACGIH, 1983).

2. American Industrial Hygiene Association. "Hygienic Guide Series" (Akron, Ohio: AIHA).
3. American Industrial Hygiene Association. "Hygienic Guide Series—Acetylene" (Akron, Ohio: AIHA, 1967).
4. Gleason, M.N., R.E. Gosselin, H.C. Hodge, and R.P. Smith. *Clinical Toxicology of Commercial Products,* 3rd Ed. (Baltimore, Maryland: The Williams and Wilkins Co., 1969).
5. Clayton, G.D. and F.E. Clayton, Eds. *Patty's Industrial Hygiene and Toxicology,* Vols. 2A–C, 3rd Ed. (New York, New York: John Wiley and Sons, 1981).
6. Colen, F.H. "A Study of the Interchangeability of Gas Detector Tubes and Pumps," *Amer. Industrial Hygiene Assoc. J.* 35, No. 11, 686 (1974).
7. American Conference of Governmental Industrial Hygienists. *Air Sampling Instruments,* 5th Ed. (Cincinnati, Ohio: ACGIH, 1978).
8. State of Oregon Occupational Health Regulations, Appendix B, Section 22–1017(A).
9. OSHA General Industry Safety and Health Regulation, U.S. Code of Federal Regulations, Title 29, Chapter XVII, Part 1910.1000 (Revised 1981).

CHAPTER 5

How Local Exhaust Systems Work

This chapter explains how local exhaust systems work. Each part of the system will be described along with the way air flows through it. An overview that explains how a ventilation system works and what physical laws govern will provide a good background before focusing on specifics like hood selection. Since air is invisible it may be difficult to visualize the movement of air through a room, into a ventilation hood, and through the ductwork. In order for a system to work properly, however, it must be designed according to the laws of fluid dynamics. As a corollary, ventilation systems that do not work properly are deficient because one or more airflow principles are violated in their design or operation.

Of course the definition of a proper or deficient ventilation system varies from one system to another. The ideal is to design a system that captures or confines all contaminants from an operation. With many hood designs, however, complete source control is not feasible or even possible because of random air currents, the velocity of the contaminant stream, or other factors. Even if total control at each hood is achieved, the level of contaminants in the workplace may still not be zero due to leaks and other uncontrolled sources.[1] For most industrial exhaust systems not handling highly toxic materials, an effective rather than an ideal system is the design goal. An effective system is one that maintains exposures below the acceptable level for the contaminant without undue installation or operating costs.[2] The acceptable level may be part of the federal or state OSHA standards, or it may be a fraction of that level depending on worker exposures to other contaminants.

Dilution exhaust systems lower the contaminant concentrations by dilution with fresh air but do not reduce or eliminate the total amount of contaminants released. In contrast, local exhaust systems capture air contaminants at or near their source before they are dispersed into the workroom air. Therefore, local exhaust is frequently the preferred method for controlling airborne concentrations of potentially hazardous materials. A typical local exhaust system consists of (Figure 5.1):

Hoods—the openings into the ventilation system where contaminants are either captured or retained by flowing air currents. The hoods are the most important part of the system—so important that the next chapter is devoted solely to the theory and practical considerations of hood selection and design. Different hoods work in different ways: some reach out and capture contaminants; others catch contaminants thrown into the hood; still others contain contaminants released inside the hood and prevent them from escaping into the workroom.

Ducts—the network of piping that connects the hoods and other system components.

Air Cleaner—a device to remove airborne materials that are carried in the exhaust air but cannot be discharged into the community environment, or to reclaim valuable materials such as silver dust or titanium dust. Air cleaners are also needed when it is safe to recirculate the exhausted air back into the plant. Air cleaners to remove both solid (particulate) and gaseous contaminants are available.

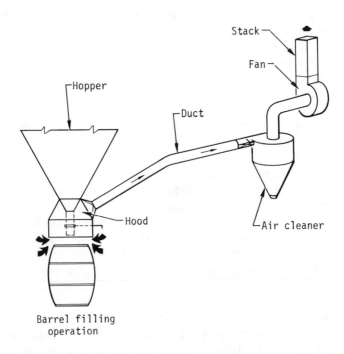

Barrel filling
operation

Figure 5.1 A typical local exhaust ventilation system consists of hoods, ducts, air cleaner, and fan.

Fan—the air-moving device that provides the energy to draw air and contaminants into the exhaust system by inducing a negative pressure or suction in the ducts leading to the hoods. The fan converts electrical power or another form of energy into negative pressure and increased air velocity.

A ventilation system is usually designed to fit existing machinery or industrial processes. A hood shape and location are chosen depending on the source of contamination. The airflow volume into each hood is then determined from reference sources or from experiments with models of the hoods when novel shapes are chosen. With this information the ducts and air cleaners can be sized. The fan size needed to draw the required amount of air while overcoming friction and other resistance can be determined. Now the system is ready for installation and testing to determine whether it meets design criteria. This book covers the individual components and overall system design. First, however, the airflow principles that explain how the system works must be understood.

PRESSURE IS THE KEY

Air starts moving because there is a difference in pressure between two points. Whether you are discussing a ventilation system or global weather patterns, air moves from an area of higher pressure to an area of lower pressure. The lower pressure in exhaust systems is generated by the fan and is called negative pressure, or suction, on the inlet side of the fan where the hoods are connected.

Pressure Represents Energy

The fan is powered by an electric motor and works by converting one type of energy (electrical) into another (pressure). The suction developed by the fan extends back through the ducts to each hood. There the suction starts room air moving into the hoods and through the system. Thus electrical energy is converted into suction (or potential energy) which in turn is changed into air movement (or kinetic energy).

The concept that pressure represents energy is important when designing or evaluating ventilation systems because the amount of energy that must be added to the system by the fan equals the energy needed to draw the proper amount of air into the hoods and also overcome the friction and turbulence in the ducts. Since pressure represents energy and is easy to visualize in a ventilation system, pressure is a convenient parameter to focus on in understanding how local exhaust systems work.

Units of Pressure

Pressure is defined as force per unit area. Familiar units of measurement are pounds per square inch (psi) or pounds per square foot (psf). If the pressure is measured on a gauge that reads zero when open to the atmosphere, the pressure is called gauge pressure (psig). If the reference is a total vacuum rather than atmospheric pressure, then the pressure is called absolute pressure (psia). For example, a device indicating absolute pressure would read 14.7 psi (normal atmospheric pressure) when open to the atmosphere at sea level. Units of psi or psf are not too useful in ventilation work because of the very small pressures involved. The fan in a typical pedestal grinder exhaust system (Figure 5.2) develops a maximum of about 0.1 psig negative pressure. Instead of using such small numbers, it is more convenient to express pressure as the height of water column that the pressure in the duct will support. Visualize the pressure exerted by a cubic foot of water (Figure 5.3). Since a cubic foot of water weighs 62.4 pounds, the pressure at the base of the cube is 62.4 psf or 0.433 psi. The 0.433 psi pressure is caused by the weight of a column of water 12 inches high. Thus 12 inches of water is equal to 0.443 psi and so 1 inch of water equals 0.04 psi. In a ventilation system, the pressure is measured directly by using a water manometer, and pressure is expressed in "inches of water" (Figure 5.4).

Types of Pressure

There are two components to the total pressure in a local exhaust system: suction, or static, pressure pulling inward on the ducts of the system;

Figure 5.2 Pedestal grinder exhaust ventilation system.

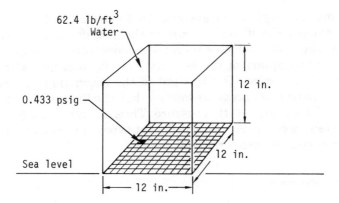

Figure 5.3 Pressure is expressed as the height of a water column that exerts the equivalent pressure. One foot of water equals 0.433 psig. (Source: Reference 3)

and velocity pressure due to air moving through the system. Since velocity pressure is more difficult to visualize than static pressure, you may want to do the following exercise as an illustration: Hold your hand palm forward outside the window of a moving car. The force you feel is due to the mass of air impacting on your hand. This force is called *velocity pressure* and its magnitude is directly proportional to the square of the air velocity. To make things more interesting, the two kinds of pressure in a ventilation system, static and velocity pressure, are related.[4] One type of pressure can be converted into the other and vice versa. Their sum is the total pressure in the system. The total pressure represents

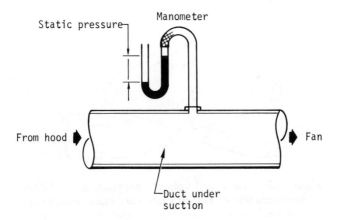

Figure 5.4 A water-filled manometer indicates pressure as "inches of water" column displaced.

the amount of energy in the system. As illustrated in Figure 5.5 by the pressure gauges, the static pressure in the duct decreases in the restriction zone where the velocity increases. Since velocity pressure is proportional to the square of the velocity, velocity pressure also increases in the restriction zone. To the right of the restriction, where velocity pressure returns to its previous lower value, the higher velocity pressure is converted back into static pressure. Thus the total amount of energy in the system is conserved. The different types of pressure, discussed in the order of ease of measurement, follow.

Static Pressure

Static pressure in a ventilation system acts to collapse the walls of the ducts on the suction side (inlet) of the fan and to burst the ducts on the discharge side. It acts equally in all directions and can be measured using a water manometer to read the bursting force on the duct walls (Figure 5.6a). Static pressure produces initial air velocity in the system and is needed to overcome the resistance to airflow caused by friction and turbulence. It represents the potential energy in the system.

Total Pressure

Total pressure is the sum of static and velocity pressures and represents the total energy in the system. It is measured by using an impact tube (Figure 5.6b) that senses both the velocity pressure in the direction of flow and the static pressure, which is exerted uniformly in all directions, inside the ducts.

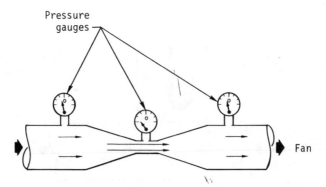

Figure 5.5 Static and velocity pressure are related. At locations where velocity pressure is increased (the restriction in the center of the figure), static pressure decreases. Reducing velocity pressure increases static pressure.

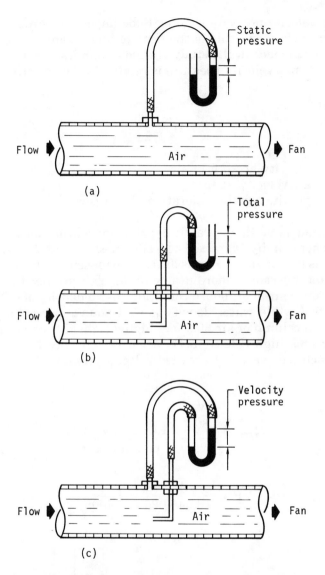

Figure 5.6   Velocity pressure is determined by subtracting static pressure from total pressure: VP = TP − SP

Velocity Pressure

Velocity pressure is exerted by air in motion and has a positive sign in the direction of airflow. It represents the kinetic energy in the system and is found by subtracting the static pressure from the total pressure (Figure 5.6c).

Since velocity pressure is going to be important through several of the following chapters, this is the time to understand it. Bernoulli, an 18th century pioneer in the study of fluid dynamics and other scientific fields, found that velocity pressure is related to the velocity of the fluid as follows:

$$H = \frac{v^2}{2g} \tag{5.1}$$

where H = velocity head, ft of air
    v = air velocity, ft/sec
    g = gravitational acceleration, 32.2 ft/sec²

This equation is of little practical value in ventilation studies since velocity is not usually expressed as feet per second and the velocity head expressed as feet of air column is difficult to measure. The velocity units are easy to convert into more usable units. Velocity head units can be converted since air at standard conditions has a density of 0.075 pounds per cubic foot while water density is 62.4 pounds per cubic foot. This means that a column of air 69.2 feet high weighs the same as a column of water 1 inch high (Figure 5.7). When velocity head is expressed in units of "inches of water" it is called velocity pressure. Restating Equation 5.1:

$$VP = \frac{\left(\frac{V}{60}\right)^2}{2g} \left(\frac{1.0 \text{ in. of water}}{69.2 \text{ ft of air}}\right) \tag{5.2}$$

$$VP = \frac{V^2}{16,040,025} = \left(\frac{V}{4005}\right)^2 \tag{5.3}$$

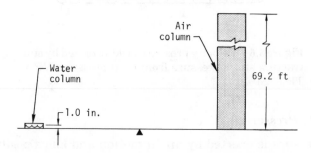

Figure 5.7   A column of water 1 inch high weighs as much as a column of air more than 69 feet high.

or

$$V = 4005 \sqrt{VP} \qquad (5.4)$$

where VP = velocity pressure, inches of water
       V = velocity, ft/min

These last two equations say that anytime you know the air velocity, you can calculate the pressure exerted by the moving air. Tables and graphs of velocity pressure as a function of velocity are also available (Figure 5.8). This relationship is important because pressure losses in the system due to friction, turbulence, and other factors described in this chapter are a function of velocity pressure which, in turn, depends on the value for duct velocity that was selected during system design. The losses can be calculated in terms of "equivalent duct velocity pressure losses" during design and then converted to units of "inches of water" using Equation 5.4. The total loss represents the pressure that must be added to the system by the fan.

Figure 5.8 shows the duct velocity pressure ranges typical of dust ventilation systems and gas or vapor systems. For example, if a dust

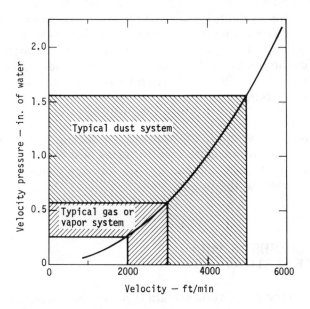

Figure 5.8 Velocity pressure as a function of velocity. Since the pressure losses are related to velocity pressure, typical ranges for systems handling dusts, and gases or vapors are shown.

system is designed with a duct velocity of 5000 ft/min, the resulting duct velocity pressure is 1.56 inches of water. Thus an air cleaner that causes a loss of 2.0 velocity pressures requires the fan to supply 3.12 inches of water suction to overcome the loss. Compare this with a vapor system designed with a duct velocity of 2000 ft/min. From Figure 5.8 the velocity pressure in this system is about 0.25 inches of water. Because of lower duct velocity, the same air cleaner will only cause 0.5 inches of water resistance in the vapor system.

Here is a summary of the importance of pressure in understanding ventilation systems:

- Pressure differences make air flow through the system.
- The total pressure in the ventilation system has two components: the velocity pressure caused by moving air, and static pressure, or suction, that starts the air moving and maintains motion.
- According to the law of conservation of energy, static pressure can be converted to velocity pressure and vice versa.
- The value of velocity pressure is proportional to the square of the velocity.
- The magnitude of resistance to airflow due to friction and turbulence depends on the square of the air velocity in the ventilation system ducts. The resistance can be expressed directly in terms of the velocity pressure in the system. It is called pressure loss or pressure drop.
- The pressure loss through the system is analogous to the energy that must be added to the system by the fan to move the air.
- The size of the fan needed to make the system work properly is calculated from the airflow needed and the pressure loss through the ventilation system.

With this background, the flow of air through each part of the ventilation system can be explained in terms of pressure loss. Pressure loss is important since it determines the fan size needed and also the power costs to operate the fan.

PRESSURE LOSSES AS AIR ENTERS THE SYSTEM

Since air enters the ventilation system through the hoods, this is a logical place to start analyzing the pressure losses. Two types of pressure loss occur.

Acceleration Loss

Air must be accelerated from the random low velocities existing in the workroom up to the duct velocity. If the room air is practically motionless, it has a velocity pressure or kinetic energy of zero. Once in the ventilation system, its velocity pressure depends on the duct air velocity according to Equation 5.3. So the equivalent of "one velocity pressure" of energy has been expended to accelerate the air. Even if the ventilation system were ideal and had no friction or turbulence, the acceleration loss would still occur. How big is this acceleration loss in typical ventilation systems? Figure 5.8 illustrates the relationship between velocity and velocity pressure. Duct velocities for gas or vapor control ventilation systems usually range from 2000 to 3000 ft/min with corresponding velocity pressures of 0.25 to 0.56 inches of water. Just getting the air moving requires this much energy.

Hood Entry Loss

As the air is accelerated into the hood, the hood shape causes turbulence that interferes with smooth airflow. The turbulence converts some of the velocity pressure into heat; this results in the loss of useful energy from the system.

The hood entry loss can be explained by the formation and dissolution of a *vena contracta,* which means a contraction of the airstream diameter. As air is pulled into a duct opening from the area outside the duct (Figure 5.9), the individual airstreams converge to a higher velocity

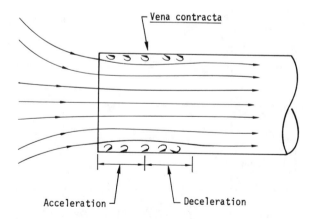

Figure 5.9   Air entering a hood or open duct forms a *vena contracta,* caused by converging airstreams.

stream that does not completely fill the duct. The point where the air-stream diameter is smallest is called the *vena contracta*. At this point the airstream diameter is about 88% of the duct diameter.[5] The velocity at this point is higher than the average duct velocity in the rest of the system since the same amount of air is passing through a smaller cross-sectional area. Thus, according to the law of conservation of energy, velocity pressure is increased along with a corresponding decrease in static pressure. As the air passes the *vena contracta* it slows down, and the airstream again fills the duct. The deceleration is accompanied by turbulence, which results in some energy loss. This loss is called the hood entry loss.

Hood Entry Loss Factor (F_h)

The hood entry loss, expressed as "equivalent duct velocity pressure loss," is called the *hood entry loss factor* (F_h) and has been measured experimentally for many hood shapes (Figure 5.10). As expected, a hood shape that does not cause much turbulence has a lower entry loss factor than a hood that causes formation of a pronounced *vena contracta*. For example, a plain duct opening has a hood entry loss factor of about 90

Hood type	Shape	Hood entry loss factor, fraction of duct VP (F_h)	Typical hood entry loss, in. of water	
			Dust system hood	Vapor system hood
Unflanged	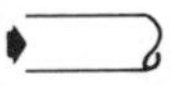	0.90	0.5-1.4	0.2-0.5
Flanged	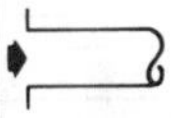	0.50	0.3-0.8	0.1-0.3
Rounded	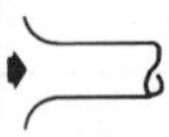	0.03	0-0.1	Negli-gible
45° taper	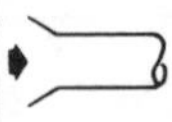	0.10	0.1-0.2	0-0.1
Slot	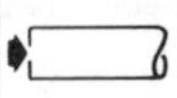	1.78 of slot VP	0.5-2.0	0.5-2.0

Figure 5.10   The effect of hood shape on the hood entry loss factor and typical hood entry loss values.

percent of the duct velocity pressure while a bell-shaped entry reduces the turbulence so that the loss factor is just 3% of the duct velocity pressure.

At the other end of the scale, a narrow slot hood, often used for proper air velocity distribution on electroplating tanks, causes such a severe *vena contracta* that the entry loss factor is almost twice the slot velocity pressure. When the slot leads into a plenum chamber and then into the main duct, as illustrated in Figure 6.24c, there is a second entry loss as the air enters the duct. This loss occurs in addition to the slot entry loss equal to 1.78 times the slot velocity pressure. Thus the slot hood has a relatively large hood entry loss.

Role of Duct Velocity

The hood entry loss expressed as "inches of water" is calculated by multiplying F_h by the duct or slot velocity pressure. Typical values are tabulated in Figure 5.10. The hood entry losses differ for dust and vapor control systems. This is because the duct velocity, and so the duct velocity pressure, is higher in dust systems to avoid dust settling in the ducts. The hood entry losses for both types of systems range from almost zero to 2 or 3 inches of water. The hood entry loss factors for many different hood shapes can be found in the *Industrial Ventilation Manual* published by the American Conference of Governmental Industrial Hygienists (ACGIH).[6]

Hood Static Pressure

This is a good time to tie static and velocity pressures together. So far the hood acceleration loss and the hood entry loss have been described in terms of duct velocity pressure. This is a convenient term because the duct velocity in a system is chosen at the start of design for a new system or can easily be measured in existing systems. But for a hood to operate properly the fan must generate enough suction or static pressure (view it as potential energy) in the duct near the hood to overcome both the acceleration loss and the hood entry loss. At the same time the fan must draw the correct amount of air into the hood. This quantity of static pressure generated by the fan is called the *hood static pressure* and is easily measured using a water manometer (Figure 5.11). Mathematically, hood static pressure is:

$$
\begin{aligned}
SP_h &= \text{Acceleration Loss} + \text{Hood Entry Loss} \\
&= (1.0 \times \text{duct velocity pressure}) \\
&\quad + (F_h \times \text{duct velocity pressure}) \\
&= 1.0 \, VP_d + F_h VP_d \\
SP_h &= (1.0 + F_h) \, VP_d
\end{aligned}
\tag{5.5}
$$

Figure 5.11 Hood static pressure, measured in the duct near the hood, represents the suction or potential energy available to draw air into the hood.

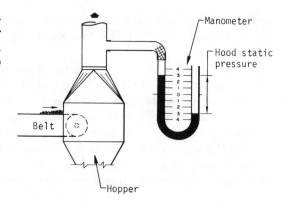

where SP_h = hood static pressure, inches of water
F_h = hood entry loss factor, dimensionless
VP_d = duct velocity pressure, inches of water

If the duct velocity is known, the duct velocity pressure can be found by using Figure 5.8. Once a ventilation system is installed and operating properly, the hood static pressure measurement is an easy way to periodically check the amount of air flowing into the hood.

Example: The barrel-filling hood in Figure 5.1 has a hood entry loss factor of 0.25 VP. Measurements show that the hood static pressure is 1.8 inches of water in the 4-inch-diameter duct at the hood. Estimate the airflow into the barrel-filling station hood.

Answer: A summary of findings is:
$$F_h = 0.25$$
$$SP_h = 1.8 \text{ inches of water}$$
$$\text{Duct diameter} = 4 \text{ inches (Area} = 0.087 \text{ ft}^2)$$
$$\text{Find airflow (Q), ft}^3/\text{min}$$

From Equation 5.5:
$$SP_h = (1 + F_h) VP_d$$
$$1.8 = (1 + 0.25) VP_d$$

Solving for VP_d:
$$VP_d = 1.44 \text{ inches of water}$$

From Equation 5.4:
$$V = 4005 \sqrt{VP_d}$$
$$V = 4005 \sqrt{VP_d} = 4005 \sqrt{1.44} = 4806 \text{ ft/min}$$
$$\text{Airflow (Q)} = \text{velocity} \times \text{area}$$
$$Q = 4806 \text{ ft/min} \times 0.087 \text{ ft}^2$$
$$Q = 418 \text{ ft}^3/\text{min}$$

Coefficient of Entry

Another way to express hood entry loss is the *Coefficient of Entry*. Coefficient of Entry is the ratio of the actual airflow into the hood to the theoretical airflow if all hood static pressure could be converted into velocity as would be the case if $F_h = 0$. The Coefficient of Entry is:

$$Ce = \frac{Q_{actual}}{Q_{max}} \qquad (5.6)$$

where Ce = coefficient of entry, dimensionless
  Q_{max} = theoretical maximum airflow, ft³/min
  Q_{actual} = actual airflow in system, ft³/min

and can be derived from Equation 5.4:

$$V = 4005 \sqrt{VP_d}$$

Since

$$Q = V \times A \qquad (5.7)$$

Equation 5.4 can be rewritten:

$$Q = 4005 \, A \sqrt{VP_d} \qquad (5.8)$$

where Q = airflow, ft³/min
  A = duct cross-sectional area, ft²

So Equation 5.6 becomes:

$$C_e = \frac{4005 \, A \sqrt{VP_{d(actual)}}}{4005 \, A \sqrt{VP_{d(max)}}} = \frac{\sqrt{VP_{d(actual)}}}{\sqrt{VP_{d(max)}}} \qquad (5.9)$$

From a practical standpoint, $VP_{d(actual)} = VP_d$, and $VP_{d(max)}$ would occur if the hood had no entry loss. In this case Equation 5.5:

$$SP_h = (1.0 + F_h) \, VP_d$$

becomes, with $F_h = 0.0$:

$$SP_h = (1.0 + 0.0) \, VP_d = VP_{d(max)} \qquad (5.10)$$

meaning that there are no losses due to turbulence. All of the hood static pressure accelerates room air to a higher VP_d than would occur if F_h were not zero.

  Equation 5.9 can be restated:

$$C_e = \frac{\sqrt{VP_{d(actual)}}}{\sqrt{VP_{d(max)}}} = \frac{\sqrt{VP_d}}{\sqrt{SP_h}}$$

$$C_e = \frac{\sqrt{VP_d}}{\sqrt{SP_h}} \tag{5.11}$$

Equation 5.11 states that you can calculate C_e for a hood by measuring the velocity pressure (see Figure 5.6c) and hood static pressure (see Figure 5.11).

Once the value of C_e is determined for a hood, the airflow (Q) through the hood can be calculated from hood static pressure readings by rearranging Equation 5.11:

$$C_e = \frac{\sqrt{VP_d}}{\sqrt{SP_h}}$$

so that

$$\sqrt{VP_d} = C_e \sqrt{SP_h}$$

and substituting in Equation 5.8:

$$Q = 4005 \text{ A } \sqrt{VP_d}$$

$$Q = 4005 \text{ A } C_e \sqrt{SP_h} \tag{5.12}$$

Equation 5.12 is convenient to use in field evaluations of hood performance since SP_h is easy to measure. Table 5.1 lists C_e values for different hood shapes.

PRESSURE LOSSES IN THE DUCTS

Air flowing through the ductwork meets resistance in the form of friction or turbulence. Like hood entry losses, losses from turbulence can be

Table 5.1 Coefficient of Entry for Different Hood Shapes

Hood Type	Coefficient of Entry (C_e)
Unflanged	0.72
Flanged	0.82
Rounded entry	0.98
45° taper	0.92
Slot	0.60

expressed as pressure drop since they represent additional pressure that the fan must generate to make the system work properly. As with entry losses, a ventilation system designed to minimize duct losses will perform adequately with a smaller fan compared to a system with high duct losses.

Friction Losses

Air movement is always accompanied by friction where the air meets the duct surface. A term such as *duct velocity* usually means average duct velocity since there is a distinct velocity profile in the moving air (Figure 5.12). The air velocity near the duct wall is almost zero while the air at the center of the duct is traveling at speeds exceeding the average velocity. As a rule-of-thumb multiplying the centerline velocity by 0.9 will provide an estimate of the average duct velocity.[7] Since there are other random air velocities in the duct, Figure 5.12 depicts a simplified view of duct velocity profiles. The amount of friction loss is related to other system parameters as follows:

- Friction varies directly with the duct length.
- Friction varies directly with the square of the velocity.
- Friction varies directly with the roughness of the duct interior.
- Friction varies inversely with the duct diameter.

Example: An existing 4-inch-diameter duct 50 ft long is carrying 400 ft³/min of air with a measured friction loss of 4.5 inches of water. If the airflow is increased to 450 ft³/min, what is the new friction loss?

Answer: Friction varies directly with the square of the velocity. Since
$$\text{Airflow (Q)} = \text{Velocity} \times \text{Area}$$
the friction loss also varies with the square of flow rate.
$$\text{Friction loss} = 4.5 \text{ in. of water} \left(\frac{450 \text{ ft}^3/\text{min}}{400 \text{ ft}^3/\text{min}}\right)^2$$
$$= 5.7 \text{ in. of water}$$

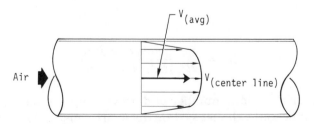

Figure 5.12 Velocity distribution in a duct. The average duct velocity equals about 90 percent of the centerline velocity.

Friction loss values for duct design are usually obtained from charts that are constructed from the general Fanning-D'Arcy equation:

$$F_f = f \frac{L}{D} \frac{v^2}{2g}$$ (5.13)

where F_f = friction loss, ft of flowing fluid
   f = friction loss coefficient, dimensionless
   L = duct length, ft
   D = duct diameter, ft
   v = velocity, ft/sec
   g = gravitational acceleration, 32.2 ft/sec²

The problem with using the general Fanning-D'Arcy equation is that the friction loss is expressed in units of "feet of air" rather than "inches of water." Also the units for duct diameter and velocity are not standard units used in ventilation work. Converting the units yields a usable equation:

$$h_f = (7.5 \times 10^{-7}) f \frac{L}{D_d} V^2$$ (5.14)

where h_f = friction loss, inches of water
   D_d = duct diameter, inches
   V = velocity, ft/min
   L = duct length, ft
   f = friction loss coefficient, dimensionless

Friction loss can be expressed in terms of duct velocity pressure using Equation 5.3.

$$VP = \left(\frac{V}{4005}\right)^2$$

and so,

$$h'_f = \frac{h_f, \text{ inches of water}}{VP, \text{ inches of water}}$$ (5.15)

where h'_f = friction loss, number of duct velocity pressures.
   Most industrial exhaust system ducts are constructed of galvanized sheet metal. Assuming round clean ducts with about 40 joints per 100 feet and standard air (density = 0.075 pounds/ft³), the friction loss coefficient f ≅ 0.27. By substituting this value of f in Equations 5.14

and 5.15, the friction losses in ducts can be calculated and plotted (Figures 5.13 and 5.14) in terms of inches of water and velocity pressure equivalents.

Turbulence Losses

Turbulence, resulting from changes in air velocity and direction, also causes pressure losses. Some loss will occur at every hood, elbow, duct enlargement, or duct junction, but the magnitude of the pressure drop depends on the efforts made during design to minimize turbulence. Sometimes the analogy is drawn between ventilation system ducts and a super highway. Air has weight and, like an automobile, tends to travel in straight lines. Thus gentle curves instead of sharp bends and gradual diameter reductions rather than abrupt contractions help to minimize congestion or turbulence. Because turbulence losses are the result of moving air, they are also called dynamic losses.

Problems with turbulence increase as a ventilation system gets bigger. When a system has several hoods and a branch duct from each hood joining into the main ducts, there are many more locations where the air speed and direction change. Compared with the other types of

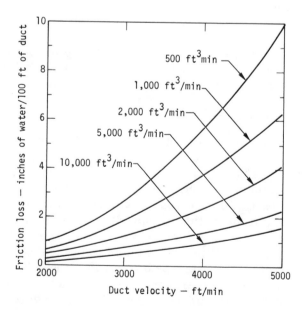

Figure 5.13 Duct friction loss, expressed in inches of water per 100 feet of duct, as a function of duct velocity for different airflow rates.

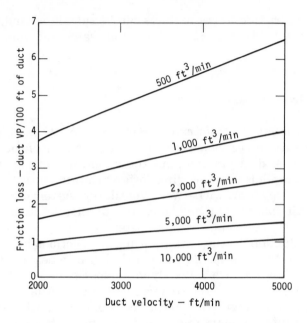

Figure 5.14 Duct friction loss, expressed in duct velocity pressure equivalents per 100 feet of duct, as a function of duct velocity for different airflow rates.

pressure losses discussed earlier, dynamic loss probably will not be a major source of pressure drop in a small, simple ventilation system. However, here are some different turbulence sources:

Abrupt Enlargements

When the cross-sectional area of a duct increases suddenly, a zone of severe turbulence occurs along the boundary of the airstream that is increasing in size (Figure 5.15a). A dead air zone forms at the enlargement and air randomly enters and leaves the zone. This random air movement causes turbulence that robs energy from the flowing airstream. The magnitude of the pressure loss in an abrupt enlargement depends on the ratio of duct diameters at the enlargement. It is always less than 1.0 VP_d. The actual loss is found from the difference between the total pressures in the system before and after the enlargement. Since the airflow pattern at the enlargement resembles a jet of air, an analysis using theoretical fluid dynamics for jet behavior produces an equation that relates pressure loss to the velocity pressure in the duct before and after the enlargement. Called the Carnot-Borda equation, it states:[5]

$$VP_{up} = VP_{dn} + h_t \qquad (5.16)$$

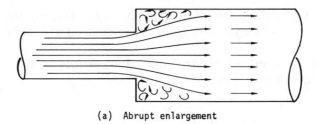

(a) Abrupt enlargement

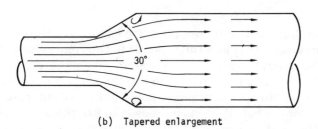

(b) Tapered enlargement

Figure 5.15 Compared with a tapered enlargement
(b), an abrupt enlargement (a) causes a severe turbulence loss.

or

$$h_t = VP_{up} - VP_{dn} \qquad (5.17)$$

where VP_{up}, VP_{dn} = duct velocity pressures immediately upstream
and downstream from the abrupt enlargement,
inches of water
h_t = turbulence pressure loss, inches of water

From Equation 5.3 (For standard air):

$$VP = \left(\frac{V}{4005}\right)^2 \text{ for standard air}$$

Equation 5.17 can be restated:

$$h_t = \left(\frac{V_{up}}{4005}\right)^2 - \left(\frac{V_{dn}}{4005}\right)^2 \qquad (5.18)$$

$$h_t = \left(\frac{V_{up} - V_{dn}}{4005}\right)^2 \qquad (5.19)$$

where V_{up}, V_{dn} = duct velocities immediately upstream and
                downstream from the abrupt enlargement, ft/min

If $V_{dn} = 0$, as when the airstream discharges into a large space where
the forward velocity is zero, then

$$h_t = \left(\frac{V_{up}}{4005}\right)^2 = VP_{up} \tag{5.20}$$

which says that the turbulence pressure loss equals the entire upstream
velocity pressure. This loss corresponds to the "one velocity pressure" of
acceleration energy added to the system when the room air was first
drawn into the hood and accelerated to the system duct velocity.

    Losses in abrupt enlargements should be of interest only from a
theoretical viewpoint. There is no reason to use them in real ventilation
systems. Gradual enlargements that reduce turbulence loss can be sub-
stituted for abrupt enlargements in almost all cases.

Gradual Enlargements

As you would expect, turbulence losses are reduced if duct enlargements
are tapered rather than abrupt (Figure 5.15b). However, reducing the
taper angle increases the length of the tapered duct section. As the
tapered section length increases, so does friction pressure loss. This continues until, for small taper angles, friction losses dominate turbulence
losses. In a gradual enlargement the minimum pressure loss, including
losses from both friction and turbulence, occurs at an included angle of
about 8°. Below 8° the total loss increases due to higher friction losses;
above 8° turbulence increases the total pressure loss. Practical considerations, such as fabrication costs or space requirements, often make an
angle larger than 8° more economical. Gradual enlargements are used
to maintain constant velocity in the main duct where additional branch
ducts join (Figure 5.16). If the main duct were not enlarged, the addition
of the branch duct would increase the main duct velocity and consequently result in higher friction and turbulence losses.

Figure 5.16 Branch ducts
should enter the main duct at a
tapered enlargement to maintain duct velocities while minimizing turbulence. A small
angle of entry also reduces turbulence losses.

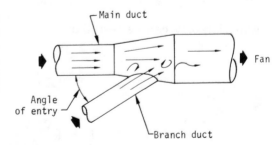

Contractions

Reductions in duct diameter are usually encountered in ventilation systems only where ducts must fit in tight spaces or pass through smaller diameter holes in walls or ceilings. Any duct contraction that is accompanied by a *vena contracta* is also accompanied by turbulence pressure loss. As in duct enlargements, abrupt contractions cause severe pressure loss and should be avoided (Figure 5.17a). A gradual contraction reduces or eliminates the *vena contracta* formation; if the included angle is 45° or less, the turbulent loss is so low that it can be neglected in pressure drop design calculations (Figure 5.17b). Even if the taper is 135°, the pressure loss is only about 25 percent of the smaller diameter duct velocity pressure.[5] There is, then, little reason to pay the price of an abrupt contraction when it is so easy to minimize this loss.

Elbows

Turbulence losses occur every time the airstream changes velocity or direction. Elbows and bends in the duct system cause these types of

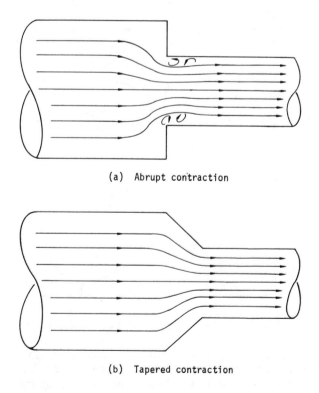

(a) Abrupt contraction

(b) Tapered contraction

Figure 5.17   A tapered duct contraction minimizes the formation of a *vena contracta*.

changes and so cause pressure losses. Hemeon [5] described the airstream that impinges on the outer wall of the elbow as a helix-shaped flow. The resulting increase in air velocity, subsequent deceleration as the air returns to normal flow, and increased friction and turbulence, are all sources of pressure loss. The magnitude of elbow losses depends on the duct velocity and the radius of the elbow curvature compared to the duct diameter. Figure 5.18 shows the losses that occur in round ducts. Wherever possible a radius of curvature at least twice the duct diameter should be used. Measurements have shown that this curvature causes 0.27 VP loss for each elbow. For a typical dust ventilation system with 3000–5000 ft/min duct velocity the loss at each elbow is 0.13–0.40 inches of water. For gas or vapor ventilation systems with 2000–3000 ft/min duct velocities the resulting loss is about 0.1 inch of water per elbow.

Branch Ducts Entering Main Ducts

Ventilation systems with more than one hood have branch ducts from each hood connecting to a main duct running to the fan. The branch duct entry causes turbulence that is a pressure loss in the system. To

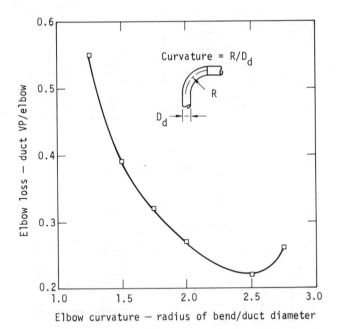

Figure 5.18   Elbow pressure loss as a function of elbow curvature. For curvatures exceeding 2.5, the friction loss due to the longer duct section at the elbow increases the overall loss. (Source: Reference 4)

minimize the turbulence, the branch duct should join the main duct at a gradual enlargement (Figure 5.16). The diameter of the main duct following the enlargement is calculated to maintain the design duct velocity, as the airflow from the branch enters the main duct. The angle of entry is also important. As expected, there is less turbulence and pressure loss with a smaller angle between the branch and main ducts (Figure 5.19). With a 15° angle only 0.3 branch duct VP is lost while for a 90° (perpendicular) entry the loss is one full velocity pressure equivalent.[6]

Other Turbulence Losses

In addition to the sources of pressure losses already covered, other ventilation system components may cause significant pressure loss. Anything that disturbs the smooth flow of air will cause turbulence losses. For example, air-cleaning devices may be a major source of pressure loss. For filters the differential pressure or pressure drop across the filter (Figure 5.20) indicates when it needs cleaning or replacing. Other air cleaners, like cyclones and electrostatic precipitators, induce turbulence to an otherwise smooth airflow pattern. The resulting pressure loss is due to this turbulence in contrast to the filter that acts as a direct airflow

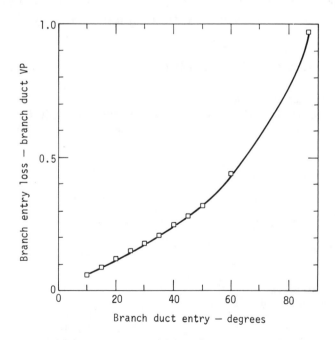

Figure 5.19 Branch duct entry loss as a function of branch duct entry angle. (Source: Reference 4)

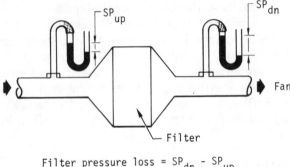

Filter pressure loss = SP$_{dn}$ - SP$_{up}$

Figure 5.20 The resistance or pressure drop across a filter indicates the need for cleaning or replacement.

resistance. Table 5.2 lists typical pressure losses for different types of air cleaners. Often the resistance of each type of air cleaner increases with increased collection efficiency.

Another serious cause of pressure loss in many ventilation systems is poor duct arrangements leading into or out of the fan. Poor fan inlet connections cause turbulence or spinning air motion at the fan preventing it from moving the air with maximum efficiency.

PRESSURE LOSSES SUMMARIZED

This chapter has described the major sources of pressure loss in a ventilation system. The information is useful since one objective in system design is to minimize the pressure losses so the smallest possible fan

Table 5.2 Typical Pressure Losses From Air Cleaners

Type	Pressure Loss,[a] Inches of Water
Filter	3–8
Cyclone	1–6
Electrostatic precipitator	0.5–1
Scrubber	
Packed tower	1–3
Wet centrifugal or	
venturi collector	2–6
Carbon adsorption bed[b]	1–10

[a]Actual pressure loss is highly dependent on collection efficiency.
[b]Depends on carbon pore size and air velocity through bed.

can be used. How important are the different sources? Table 5.3 lists typical ranges for the losses generally encountered in industrial exhaust systems. If a system has an air cleaner, the air cleaner is probably the largest source of pressure losses. The next largest source is duct friction losses followed by acceleration losses and hood entry losses. Elbows and branch duct entries cause minor losses in most systems.

PRESSURE LOSS DISTRIBUTES AIR

Pressure loss has another function in multiple-hood ventilation systems: it distributes the airflow between the different hoods and branch ducts. Suppose a second hood is added to the barrel-filling ventilation system in Figure 5.1. The new hood is on the hopper above the barrel-filling station and will be connected to the existing air cleaner and fan (Figure 5.21). The new hood needs 200 ft^3/min of inward airflow to operate properly, while the existing hood needs 400 ft^3/min. How can the system be designed to draw the correct flows through each hood?

The answer is that flow is distributed so that the suction or static pressure at the junction in each of the two branch ducts (Point A in Figure 5.21) is equal. The new duct and hood should be designed to create the same resistance between the hood and the junction with an airflow of 200 ft^3/min as occurs in the duct between the original hood and the junction with an airflow of 400 ft^3/min. Then the remaining ducts and fan can be designed to develop the proper amount of suction at point A so that each hood has the correct airflow. When the system is installed and operating, the airflow distribution will automatically

Table 5.3   Ventilation System Pressure Losses

Type of Losses	Typical Magnitude, Inches of Water	Reason
Acceleration losses	0.25–1.5	Energy needed to accelerate air to duct velocity
Hood entry losses	0.1–2.0	Turbulence as air enters hood and ducts
Duct friction losses	1.0–5.0 per 100 ft of duct	Friction as air moves through duct
Turbulence losses		Turbulence as air changes direction or velocity
Elbow	0.1–0.3 per 90° elbow	
Branch entry	0.1–0.3 per 45° entry	
Enlargements and contractions	0.1–0.5 per enlargement or contraction	
Air cleaners	0.5–10	Friction and turbulence

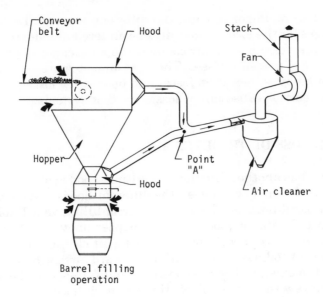

Figure 5.21 The airflow to each hood in multiple-hood systems is determined by pressure loss in the hood and branch ducts. The airflow will automatically balance so the static pressure in each branch duct at point A is equal.

balance itself until the static pressure (i.e., resistance) in both ducts is equal. If the system is poorly designed, however, the airflow will not meet design criteria in one or both hoods. Proper air distribution is achieved by choosing duct diameters that produce the correct resistance or by using adjustable dampers in some ducts to provide artificial pressure loss to balance the airflow. The design of multiple hood ventilation systems is described in Chapter 8.

FAN ADDS ENERGY TO SYSTEM

The fan provides the energy to overcome the pressure losses as air flows through the system. Electricity or another fuel powers a motor that spins the fan wheel or blades. The fan draws air in and discharges it at a higher velocity and static pressure. Several different fan types are available that are intended for use under specific velocity and static pressure conditions.

The different fan types and their operating characteristics are covered in Chapter 9. However, for all fans the quantity of air they move depends on the resistance or pressure loss in the ventilation system. If a fan is tested without any ducts and hoods connected to it, its output will be greater than when it is connected to an exhaust ventilation system. The effect of static pressure (resistance) on airflow can be plotted and is called the fan's rating or characteristic curve.

Energy consumption is also related to system airflow resistance. It takes a bigger motor to deliver the same airflow against higher resistance. Figure 5.22 illustrates electrical consumption as a function of system resistance. As long as a fan is operating within its design range, a system with only half the resistance of another system consumes roughly half the energy to move the same amount of air. Saving on operating costs is another reason to minimize pressure losses in the system.

SUMMARY

Pressure is a key concept in understanding how local exhaust systems work. Suction is the reduced pressure caused by the fan that draws the desired amount of air into the hoods and through the ducts and air cleaner. The amount of suction (static pressure) that the fan must generate for proper system performance represents the energy added to the

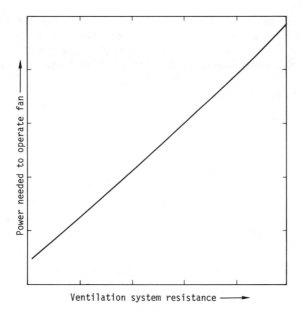

Figure 5.22   Fan power consumption is directly related to the resistance to airflow through the ventilation system at the design airflow rate.

system to overcome pressure losses from acceleration, hood entry, duct friction, and elbow turbulence as well as the resistance loss of air cleaner. The magnitude of these losses is proportional to the square of the air velocity which, when converted to pressure and expressed in units of inches of water, is called *velocity pressure*. As described in Chapter 8 (Ventilation System Design), the individual pressure losses in the system are calculated and summed to determine the fan size needed to overcome the pressure losses while moving the required volumetric airflow.

REFERENCES

1. Lynch, J.R. "Industrial Ventilation: a new look at an old problem," *Michigan's Occupational Health* 19, No. 3, 4 (1974).
2. Astleford, W. "Engineering Control of Welding Fumes," U.S. Department of Health, Education and Welfare, Publication No. (NIOSH) 75–115 (Washington, D.C.: U.S. Government Printing Office, 1974).
3. U.S. Public Health Service. "Water Supply and Plumbing Cross-Connection Hazards in Household and Community Supply Systems," Publication 957 (Washington, D.C.: U.S. Government Printing Office, 1971).
4. Federal Aviation Administration. "Pilot's Handbook of Aeronautical Knowledge" (Washington, D.C.: U.S. Government Printing Office, 1971).
5. Hemeon, W.C.L. *Plant and Process Ventilation* (New York: Industrial Press, Inc., 1963).
6. ACGIH Committee on Industrial Ventilation. *Industrial Ventilation—A Manual of Recommended Practice,* 17th Ed. (Lansing, Michigan: American Conference of Governmental Industrial Hygienists, 1982).
7. American National Standard Z 9.2–1971. "Fundamentals Governing the Design and Operation of Local Exhaust Systems" (New York: American National Standards Institute, 1972).

CHAPTER 6

Hood Selection and Design

The hood is the most important part of a ventilation system. No local exhaust system will work properly unless enough of the contaminants are retained or captured by the hoods so that the concentration of contaminants in the workroom air is below acceptable limits. Both the design and location of the hoods are crucial in determining whether a system will work. A poor hood design may prevent the ventilation system from ever working adequately. It may also result in excessive power costs as fan size and speed are increased to compensate for the initial poor hood selection.

Hood selection is an area where the health and safety professional can make a significant contribution since the keys to good hood selection are: a knowledge of airflow principles, an understanding of the plant processes, and a familiarity with employee work patterns around each process. In many plants the health and safety staff has the best overall understanding of these three areas.

A hood is any point where air is drawn into the ventilation system to capture or control contaminants. Some hoods are designed to fit around existing machinery while others are located next to the contaminant source. Figures 6.1 through 6.5 illustrate common ventilation hoods, but even a plain duct opening is called a hood if that is where air enters the system. This is because the suction needed to draw air into the ventilation system must be calculated according to the "hood entry losses" at each opening.[1] Chapter 5 discussed these losses and the fact that the suction or hood static pressure in the duct near the hood represents the system's potential energy, which is available for drawing air into the hood (Figure 5.11). Although hood static pressure is important, it is the air velocity distribution around the hood opening, and the interaction between airborne contaminants and the flowing airstreams entering the hood that determine whether the hood, and hence the whole system, will work properly. Another practical aspect is whether the hood is usable

Figure 6.1 An exhaust hood on a lathe used for trim-
ming asbestos-concrete pipe.

once it is installed. If the hood gets in the workers' way too often, it will
probably be moved out of the way and forgotten. If a sheet metal hood
is in the path of an overhead crane or forklift truck, it may not last very
long.

Figure 6.2 An enclosing hood on a waste
bag used for disposal of empty toxic dust
containers.

Figure 6.3 Exhaust hoods for buffing and wire brush wheels.

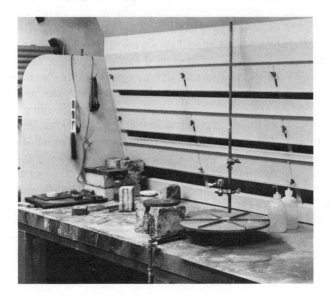

Figure 6.4 Slot hoods to distribute airflow across a soldering bench.

Figure 6.5 Close capture hood on lathe used for smoothing the exterior of asbestos-concrete pipe.

Luckily once you understand the fundamentals of hood selection, there is a ready reference source for specific hood designs. The *Industrial Ventilation Manual*[2] published by the American Conference of Governmental Industrial Hygienists (ACGIH) has over 100 design plates showing layout, design parameters, and airflow recommendations for different hoods. For example, the *Manual* contains two different hoods for welding operations: a welding bench and a portable hood (Figure 6.6). Since either will work, the choice depends on the plant layout and the welding jobs. Once the type of welding hood and its dimensions are chosen, the airflow can be calculated and the entry loss and minimum duct velocity criteria used in duct and fan design can be read from the drawing. The *Manual* recommendations are based on trial-and-error experience with hoods that work. The National Institute for Occupational Safety and Health (NIOSH) also sponsors research to check hood design criteria for welding hoods, grinding hoods, and open surface tank hoods. [3–7] In general the research findings show that the *Manual* design recommendations are adequate.[6] Refinements in design standards are incorporated into new editions of the *Manual* but one finding of a NIOSH study deserves special mention. This study found that some hood designs do not provide adequate protection for processes involving highly toxic materials.[4] For this reason Chapter 10 of this book covers ventilation systems for highly toxic materials.

HOOD TYPES

There are three major types of hoods (Figure 6.7), each working on a different principle:

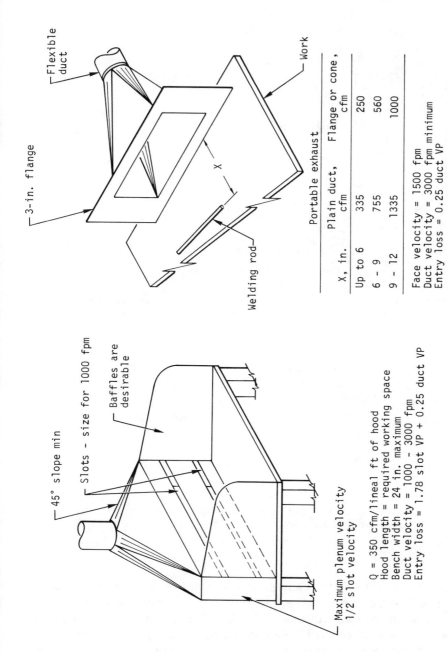

Figure 6.6 Welding hood design recommendations from ACGIH *Manual*. Similar information is available for 100 other types of hoods.

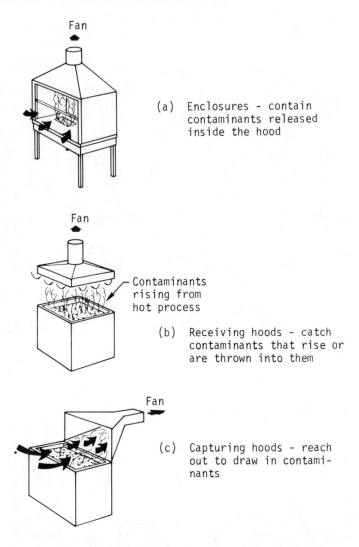

Fan

(a) Enclosures - contain
contaminants released
inside the hood

Fan

Contaminants
rising from
hot process

(b) Receiving hoods - catch
contaminants that rise or
are thrown into them

Fan

(c) Capturing hoods - reach
out to draw in contami-
nants

Figure 6.7 The three major hood types: (a) enclosures,
(b) receiving hoods, and (c) capturing hoods.

• *Enclosures*—Hoods that surround the contaminant sources as much
as possible. Contaminants are kept inside the enclosure by air flowing
in through openings in the enclosure. The quantity of air required
for contaminant control is calculated by multiplying the inward air
velocity needed to prevent escape, by the area of doorways and other
openings into the enclosure. The more complete the enclosure, the
less airflow is needed for control. Employees generally do not work

inside enclosures while contaminants are being generated, although they may reach into the enclosure as long as they do not breathe contaminated air. The design should distribute air within the enclosure to prevent accumulation of explosive or flammable vapor concentrations. If workers without protective respiratory equipment must enter the enclosure soon after the process stops, the airflow and distribution within the enclosure must be sufficient to quickly reduce airborne contaminants to acceptable levels.[7]

• *Receiving Hoods*—Some processes "throw" a stream of contaminants in a specific direction. For example, a furnace emits a hot stream of air and gases that rises above the unit. A grinder throws a stream of material tangentially from the point of contact between the wheel and workpiece. The ideal hood for this type of process is one that is positioned so it catches the contaminants thrown at it (Figure 6.7b). The airflow requirements for a receiving hood are based on the volume of contaminated air coming at it. The physical size of the hood depends on the size of the airstream. A major limitation to the use of receiving hoods is that gases, vapors, and the very small particles that can be inhaled and retained in the human respiratory system do not travel very far in air unless carried by moving air. For example, the visible chips and sawdust thrown from a power saw are not a health hazard since they are too big to remain airborne. The dust cloud that hangs in the air is made up of very fine particles each too small to see by itself. This means that receiving hoods are not very useful for health protection ventilation systems unless the process emits quantities of hot air or air with sufficient velocity to carry the respirable contaminants into the hood.

• *Capturing Hoods*—Hoods that "reach out" to capture contaminants in the workroom air (Figure 6.7c). Airflow into the hood is calculated to generate sufficient capture velocity in the air space in front of the hood. The needed capture velocity depends on the amount and motion of contaminants and contaminated air. This hood is widely used since it can be placed alongside the contaminant source rather than surrounding it as with an enclosure. The primary disadvantage is that large air volumes may be needed to generate an adequate capture velocity at the contaminant source. A second disadvantage is that the reach of most capturing hoods is limited to about 2 ft from the hood opening.

Each type of hood will be covered more fully later in this chapter. But even with these brief definitions you can see that the first question in hood evaluation or design is "what type of hood is it?" An enclosure with a relatively low airflow will contain contaminants released inside the hood. But the hood will probably not capture any contaminants released outside the hood because the inward air velocity is too low

(Figure 6.8). To control the emissions in front of the hood, the airflow must be increased or a capturing hood installed. Likewise, installing a canopy hood designed according to receiving hood criteria over a solvent cleaner at room temperature will not reduce mist and vapor emissions into the workroom. This is because the contaminants do not move into the hood under their own motion (Figure 6.9). A capturing hood is needed to reach outside and draw in the contaminants. A lot of air is needed to capture contaminants with a hood designed as a receiving hood, as shown in Figure 6.9.

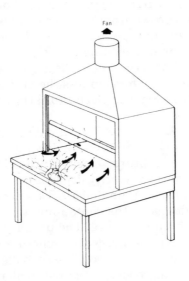

Figure 6.8   An enclosure, designed to prevent escape of contaminants released inside the enclosure may not capture contaminants generated outside of the enclosure.

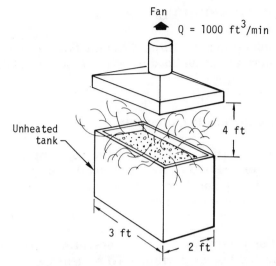

(a) Canopy designed as receiving hood
     will not control vapors from
     unheated tank

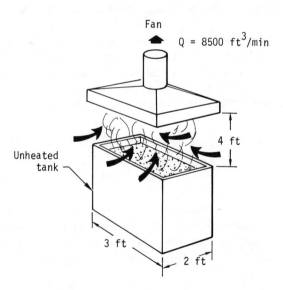

(b) Airflow must be increased 850% to use canopy
     as a capturing hood over an unheated process

Figure 6.9 A capturing hood requires sufficient
airflow to "reach" outside of the hood to capture
contaminants.

HOOD SELECTION GUIDELINES

Once you understand the three different types of hoods and how each one works, here are some hood selection guidelines (Table 6.1):[8]

Minimize Airflow Requirements

The primary objective is to control the contaminants with minimum airflow. Probably no other phase of hood design gives as great a dollar savings in the long run from reduced operating costs. These ideas can help reduce airflow requirements:

- As much as possible, enclose the operation with a ventilated enclosure, side baffles, or curtains. This measure helps to contain the material and to minimize the effect of room air currents.
- When using a capturing hood or receiving hood, locate the hood as close to the contaminant source as possible.
- Reduce the amount of contaminants generated or released from the process.
- Design the hood for good air distribution into the hood openings. In this way all the air drawn into the hood helps to control contaminants. Avoid designs where velocities through some openings have to be very high in order to develop the minimum acceptable velocity through other openings or parts of the hood.

Protect Workers' Breathing Zone

Remember that the reason for most ventilation systems is to prevent exposure to contaminants by inhalation. For this reason the hood should be located so contaminants are never drawn through the workers' breathing zone. This is especially important where workers lean over an operation such as an open surface tank or welding bench.

Table 6.1   Hood Selection Guidelines

Minimize airflow requirements.
Protect worker's breathing zone.
Follow design recommendations.
Make the hood usable by workers.
Avoid common hood selection fallacies.

Follow Design Criteria

To find the best way to control the contaminants, use applicable OSHA standards or the design criteria in the ACGIH *Manual*. Remember that most hood design recommendations do not take into account crossdrafts in the workroom that interfere with hood operation. Strong crossdrafts can easily reduce a capturing hood's effectiveness by 75% even if the hood design meets ACGIH recommendations.[5] Also keep in mind that the standard hood designs may not be adequate for highly toxic materials.

Make the Hood Usable

Design the hood so that it causes minimum interference with workers. Access doors in enclosures that must be opened and closed are often left open; capturing hoods that are too close to the process for the workers' convenience are disassembled and removed. Initial savings in airflow evaporate if the system is not used properly. Of course, the hoods should never increase mechanical injury hazards by interfering with worker movement around the machinery.

Avoid Common Fallacies

There are two common fallacies about hood design. The first is that hoods draw air from a significant distance away from the hood opening and can therefore control contaminants released some distance away. It is easy to confuse a fan's ability to blow a jet of air with its ability to draw air into a hood (Figure 6.10). The discharge flow forms a jet. The average air velocity at any point along the jet is proportional to the cross-sectional area of the airstream. The air drawn into the fan inlet duct is from a spherical volume around the duct opening; the inlet velocity decreases sharply within a single duct diameter from the opening. Even a well-designed capturing hood usually can generate an adequate capture velocity no more than two feet from the hood opening. Beyond this distance random drafts and other air currents disperse the contaminants in the room.

The second fallacy is that heavier-than-air vapors tend to settle to the workroom floor and can be collected by a hood located there. The truth is that for the small amounts of vapor in contaminated air (1000 ppm means 1000 parts of contaminants plus 999,000 parts of air), the resulting density of the mixture is so close to that of air that random air currents disperse the materials throughout the room.

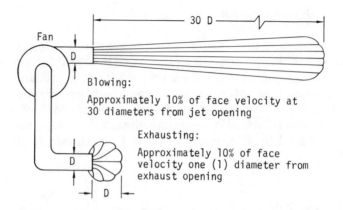

Figure 6.10 Confusing a fan's ability to blow an air jet a long distance with its capability to pull in exhaust air is a common reason why some hoods do not work.

Example: What is the density of a 1000-ppm acetone-in-air mixture? (Air = 1.0)

Answer: Density of acetone vapor = 2.0[9]
    Density of air = 1.0

$$\text{Relative Density} = \frac{1000\ (2.0) + 999,000\ (1.0)}{1,000,000} = 1.001$$

Applying Selection Guidelines

These five hood selection criteria can be applied to almost every hood design task. For example, most people are familiar with laboratory fume hoods (Figure 6.11). These hoods are found in almost every school, hospital, and industrial chemistry laboratory. The first thing to notice is that it is an enclosure designed to prevent the escape of vapors and fumes by an inward flow of air. On some hoods the sliding door or sash may have to be partially closed to generate enough velocity to contain contaminants. However, any contaminant released with a high velocity directed outward into the room will escape from the hood since the inward air velocity is not high enough to contain it. Keeping the sliding doors closed whenever possible reduces the chance for contaminant release. Also a large volume of evaporating solvent or smoke from a fire may exceed the fan's exhaust capacity, resulting in spillage into the room.

From the outside the laboratory hood resembles a box with an exhaust duct at the top leading to a fan (Figure 6.12a). If this were the case the air distribution through the hood door would be poor; probably 85% of the air would pass through the upper half of the hood door opening and the total airflow would have to be excessive in order to achieve the

Figure 6.11   A laboratory exhaust hood. *(Courtesy Kewaunee Scientific Equipment Corp.)*

needed minimum velocity at the bottom. Also there would be a dead air zone at the back of the hood counter where contaminants are generated. What you may not see is that well-designed laboratory hoods have slots or other openings along the back. These openings lead to a plenum chamber that distributes the air (Figure 6.12b) and causes it to flow horizontally into the hood with fairly uniform velocity at all points in the doorway. Since the hood's efficiency depends on maintaining some minimum inward velocity at all points in the opening, a uniform airflow significantly reduces the total airflow required. For example, to get a minimum inward velocity of 80 ft/min at any point in the doorway, about four times more total airflow would be needed for the exhausted box hood shown in Figure 6.12a than the well-designed laboratory hood (Figure 6.12b).

Laboratory hood design can get more complicated. In both laboratory hoods illustrated, all the exhausted air is taken from the laboratory. This can be expensive if heated or air-conditioned room air is continuously exhausted. Modern laboratory hoods are available with bypasses (Figure 6.13) to exhaust outdoor air entering through a separate duct

Figure 6.12   An enclosure with only a top exhaust outlet (a) draws most of the air through the top portion of the door opening. Very little air passes through the lower part of the opening. A well-designed hood (b) has slots at the rear to distribute air across the door opening.

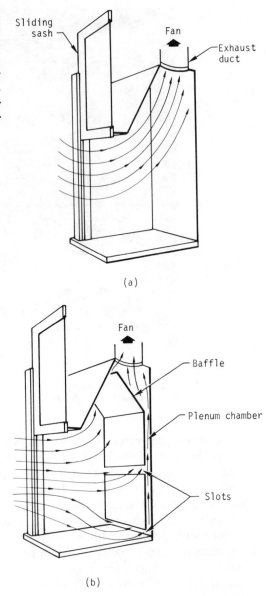

(a)

(b)

when the sash is closed. This reduces the amount of room air exhausted. Another type of hood, called a supplied-air laboratory hood, channels outdoor air through a grill into the hood face. This reduces the quantity of room air exhausted by up to 70%.

What face velocity is needed for a laboratory hood? That depends primarily on the toxicity of contaminants as well as the airflow needed

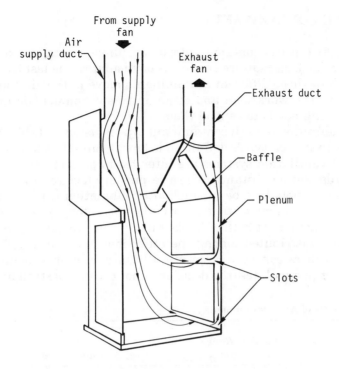

Figure 6.13 Modern laboratory hoods often have a bypass to exhaust outside air when the sash is closed. This reduces the amount of room air exhausted.

to remove explosive or flammable gases and vapors. Not too long ago 50–75 ft/min was considered adequate; current ACGIH recommendations are 100–150 ft/min[2], while OSHA requires 150 ft/min for hoods where carcinogens are handled.[10] Recent studies of breathing zone "exposures" to mannequins positioned in front of laboratory hoods during release of a tracer gas inside the hood indicate that high face velocities (> 100 ft/min) may not be beneficial in reducing exposures. In some cases higher face velocities cause eddy currents in front of the mannequin that actually draw some tracer gas out of the hood and into the mannequin's breathing zone.[18, 19] These studies also emphasized that good room conditions (sufficient replacement air, supply outlets located and designed to avoid drafts, and protection against disruptive air currents from open doors and passing workers near the hoods) are vital to laboratory hood performance.[20]

This discussion of laboratory hoods is included to illustrate the major hood selection concepts. Choose the best hood type and the minimum velocity to control the contaminant. Then look for ways to reduce airflow requirements.

AIRBORNE CONTAMINANTS

There are six types of contaminants: dusts, fumes, mists, smoke, vapors, and gases. The first four are classed as particulates; the last two are not. Knowing how these different contaminants move in the air and become dispersed in the workroom and community environment is important when selecting hoods to control them.

The definition of each contaminant type is listed in Table 6.2. More important that each type's definition is how it can affect human health. Almost all ventilation systems are intended to protect against respiratory hazards. For a substance to be a respiratory hazard, it must remain airborne long enough to be inhaled; after inhalation it must penetrate to the lungs or other susceptible parts of the respiratory system. Obviously gases and vapors meet both of these criteria since they are of molecular size and are distributed among the other molecules in air. Therefore, gases and vapors can be a respiratory hazard in high enough concentrations. For particulates, the decision is not always as straightforward.

Table 6.2 Types of Air Contaminants

Airborne Material	Size Range, μm	Characteristics
Dust (Airborne)	0.1–30.0	Generated by pulverization or crushing of solids. Typical examples are rock, metal, wood, and coal dust. Particles may be up to 300–400 μm but those above 20–30 μm usually do not remain airborne.
Fumes	0.001–1.0	Small solid particles created by condensation from vapor state, especially volatized metals as in welding. Fumes tend to coalesce into larger particles as the small fume particles collide.
Mists	0.01–10.0	Suspended liquid particles formed by condensation from gaseous state or by dispersion of liquids. Mists occur above open surface electroplating tanks.
Smokes	0.01–1.0	Aerosol mixture from incomplete combustion of organic matter. This size range does not include fly ash.
Vapors	0.005	Gaseous forms of materials that are liquids or solids at room temperature. Many solvents generate vapors.
Gases	0.0005	Materials that do not usually exist as solids or liquids at room temperature, such as carbon monoxide and ammonia. Under sufficient pressure and/or low temperature they can be changed into liquids or solids.

Source: Reference 11.

Particulates can be divided into two classes depending on their aerodynamic size. The aerodynamic size is the equivalent diameter of a sphere with the density of water that behaves like that particulate in air. It is used because different density and different shaped particles of the same diameter move differently in air. The two classes of particulates are:

- Particles or droplets larger than 30–100 μm that settle out rapidly and do not represent an airborne hazard. If these materials are generated with an initial velocity, controlling them is often easiest with a nearby receiving hood. The hood should be located so that the materials' trajectory carries them into the hood opening. Without proper ventilation, large particles settle on floors, roof beams, and machinery. Large droplets may also coat walls and make surfaces slippery or sticky. Housekeeping can be a real problem if large quantities of dust are produced, but in general these larger particulates do not present a health hazard.
- Fine particles or small droplets that do remain airborne for long periods. Their aerodynamic size is less than 50 μm and usually less than 30 μm. These small particulates are often generated along with the larger particulates especially in grinding operations. Very small particulates meet significant resistance as they move through air. Both their settling velocity and the distance they travel through air, even with high initial velocity, are low (Table 6.3). They have almost no power of motion independent of the surrounding air. So particulates of industrial hygiene significance are controlled by controlling the air in which they are suspended.

Although very small particulates remain airborne indefinitely, not all pose the same degree of respiratory hazard. First, the toxicity, or ability to do harm, varies greatly with different materials. Some compounds are very toxic while others are relatively inert and are considered "nuisance" particulates. Second, only specific size ranges of

Table 6.3   Horizontal Stopping Distance and Settling Time for Sawdust[a]

Particle Size, μm	Horizontal Stopping Distance,[b] in.	Five-Foot Settling Time, min
10	2.53	52.4
5	0.63	210.0
3	0.23	588.0
2	0.10	1300.0
1	0.03	4520.0

[a]Particle density of 10 lb/ft³.
[b]Initial velocity = 4300 ft/sec.
Source: Reference 12.

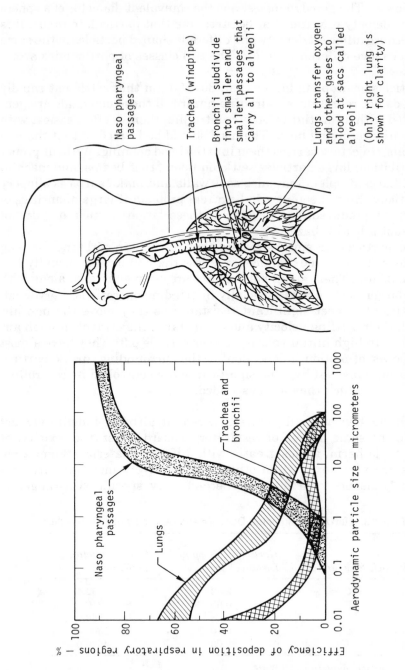

Figure 6.14 The site of deposition within the respiratory tract for airborne particles depends on their aerodynamic size. (Source: Reference 13)

particulates penetrate into the different parts of the respiratory tract. Inhaled particulates are deposited in different regions of the respiratory tract depending on their aerodynamic size. Figure 6.14 illustrates that only particles 10 μm or smaller are deposited in the lungs. Particulates larger than 10 μm are deposited in the nasopharyngeal region, removed to the throat by hair cells in the passage linings, and swallowed.[13]

Even though the size distribution of particulates reaching and depositing in the lung is known, this information is not too useful in designing ventilation systems to capture these particles. Most OSHA standards and Threshold Limit Values[15] refer to the total airborne level of contaminant, not just the portion within the respirable size range. The reason is that there was no easy way to differentiate between the respirable and nonrespirable particulates when much of the environmental data behind current standards were collected. Today small cyclone separators for air samplers are available. They separate and discard the larger nonrespirable particles according to the respirable size distribution in Table 6.4.

Controlling Airborne Contaminants

Don't be confused by this discussion of particulates and respirable or nonrespirable size particles. The main thing to keep in mind is that the contaminants capable of causing a respirable hazard move with the air around them; they have no independent action of their own. Gases (except the very light ones such as hydrogen), vapors, fumes, smoke, mists, and fine dust all move with the air currents. These contaminants are controlled by controlling the air around them. The particulates that are too large to remain airborne are the ones that can move by themselves. They do not usually pose a potential respiratory tract hazard. Controlling

Table 6.4   Respirable Dust Air Sampler Sizing Characteristics

Particle Size (aerodynamic diameter) [a] *μm*	*Percent Passing Size Selector (to be collected as respirable fraction of dust)*
2	90
2.5	75
3.5	50
5.0	25
10	0

[a]Assumes unit density sphere.
Source: Reference 15.

them, perhaps with a receiving hood, may not control the smaller particulates or vapors released from the process.

It is also important to reduce factors in the workroom or process that interfere with proper hood performance such as:

- Minimize external air currents that disperse contaminants in the workroom. Sources of air currents include open windows, heaters, or area cooling fans. Space heaters are a particularly bad source of air currents in many plants. Air currents that cannot be eliminated should be deflected by baffles around the process. Figure 6.15 illustrates that one study found a 100% increase in airborne contaminants near an open surface plating tank due to a 60 ft/min draft from a window fan blowing across the tank.[5]
- Eliminate dispersive forces from the machinery or process that help to spread the contaminants. Examples include the compressed air exhaust from pneumatic tools, long free-fall paths in dust or powder handling systems and excessive machinery vibrations that increase dust releases.

HOOD DESIGN PRINCIPLES

Earlier in this chapter three different hood types were defined: enclosures, receiving hoods and capturing hoods. Deciding which type of hood is needed to control a specific source is the first step in hood design. When evaluating existing ventilation systems you have to determine which type each hood is (or is supposed to be) so you can estimate airflow requirements. This section discusses the design or selection principles for each type of hood. These principles are general and are not detailed hood design parameters for different industrial processes. The idea is simply to help you understand how each one works. The ACGIH *Manual* is still the best source of this information.

This section does not discuss how the hood design affects the ducts, fan, and rest of the ventilation system. This was outlined in Chapter 5. Remember as you design hoods that any feature in the hood that increases turbulence or other airflow resistance, such as sharp corners or high entry velocities, means that a larger fan will be needed to pull the desired amount of air in through the hood as compared to a hood without as much resistance.

Although they are often the hardest to design, capturing hoods are discussed first since they are widely used for health protection systems. Enclosures are covered next, and finally receiving hoods. Canopy hoods are discussed as a special case of receiving hoods.

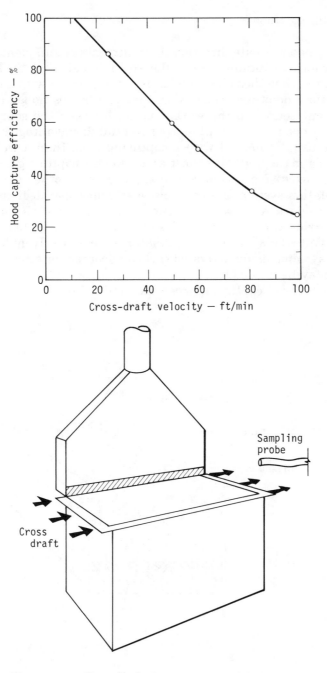

Figure 6.15 Crossdrafts have a severe impact on a capturing hood's efficiency in controlling contaminants.

Capturing Hoods

Capturing hoods create directional air currents of sufficiently high velocity to capture contaminants in the workroom air near the hood. A big advantage is that these hoods usually interfere less with the work operations than do other hoods. Also, they can be positioned close to the contaminant source so the worker is not between the source and the ventilating hood. The following are potential disadvantages: more total airflow is usually needed with a capturing hood than a well-designed enclosure; and the high air velocities around the capturing hood opening may cause excessive loss of solvents, powders, or other materials into the ventilation system. For example, a solvent vapor degreaser (Figure 6.16) cleans parts as solvent vapors condense on the parts and the liquid solvent carries away the grease. A slot exhaust installed to reduce vapor emissions into the workroom may result in excessive solvent loss. Solvent recovery systems using activated carbon adsorbers or other solvent loss controls may be needed.[16]

Four types of capturing hoods are used today:

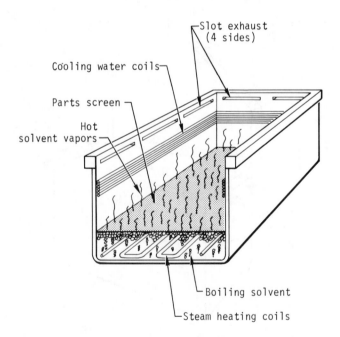

Figure 6.16 A slot hood around a solvent vapor degreaser may cause excessive solvent loss. A solvent recovery system or a control method other than ventilation should be investigated.

- *Side Draft Capturing Hoods*—The popular welding hoods or side shakeout hoods (Figure 6.17a). The hood may be movable or stationary, but the idea is to get the hood as close as possible to the source of contamination. When the distance from the hood to the source has been determined and the face area (or opening) of the hood has been selected according to source size, then the airflow into the hood must be calculated to give adequate capture velocity at the farthest point of contaminant generation.

- *Slot Hoods*—Used to ventilate narrow open surface tanks in electroplating shops or other similar applications (Figure 6.17b). The slot is a narrow opening along the tank edge leading into a plenum chamber under suction. The slot provides resistance to distribute the air along the entire length of the tank. It may allow the hood to reach out farther with lower airflow than a hood without a slot. As a rule-of-thumb, two feet is the maximum reach of a slot hood. Slots along opposite edges of the tank increase allowable tank width to four feet. Above four feet, push-pull systems (Figure 6.18) may be needed to

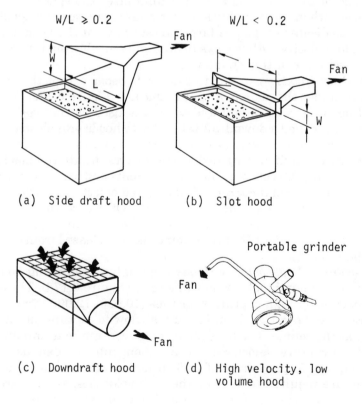

Figure 6.17   Four types of capturing hoods.

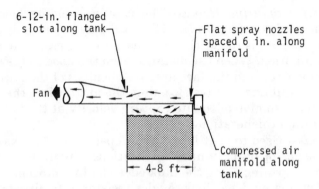

Figure 6.18 Push-pull ventilation systems are used on wide tanks since the reach of a slot hood is usually about 2 feet.

keep the fan size and power costs within acceptable limits. The advantage of the slot hood is that it minimizes the space occupied by the hood when air distribution over a large surface is needed. However, the disadvantage is a large pressure loss at the hood due to the high slot velocity. Other ways to distribute air flowing into hoods are discussed later in this chapter.

- *Downdraft Hoods*—Similar to side draft hoods except that they are located under the source of contamination (Figure 6.17c). When the hood opening (with a suitable screen) is used as the work surface, the hood is called a downdraft table. These hoods provide good control when working with solvents close to the table surface and for some soldering or torch cutting tasks. Two factors negate the hood's effectiveness: large objects may block air currents and permit contaminant escape; and thermal currents from heated objects can carry contaminants upward out of the hood's reach. Of course the airflow into downdraft tables must be designed to develop the needed capture velocity at the height that contaminants are released from operations on the table.

- *Low-volume, High-velocity Hoods*—Specialized capturing hoods that fit on power hand tools or machine tools (Figure 6.17d). These hoods use extremely high capture velocities (10,000–12,000 ft/min slot velocities are typical) and a minimum airflow to capture contaminants right at the source. The hood shape helps contain contaminants until the high capture velocity can draw them into the exhaust system. Suction (static pressures) of 6–8 inches of *mercury,* or 7–9 feet of water, are required to develop the high velocities, as compared with the several inches of *water* required with other types of hoods. Although they require a small turbine compressor rather than an exhaust fan, these low-volume, high-velocity hood systems are popular where the cost of air conditioning or heating the replacement air is high or other hoods will not work.

Overhead canopies (Figure 6.9) are not considered capturing hoods because the air velocities outside the hood are not sufficient to capture contaminants except at very high exhaust flow rates. Canopies are useful as receiving hoods for controlling hot gases or vapors rising into the canopy. Remember, it is almost never economical to use a canopy as a capturing hood.

Two Key Items: Velocity and Distribution

The following criteria govern whether or not a capturing hood is going to work properly: the hood must develop adequate capture velocity to capture the contaminants: the capture velocity must be distributed in such a way that enough contaminants are collected to reduce worker exposures to acceptable levels.

Perhaps the easiest way to visualize the operation of a capturing hood is to picture the hood as creating an imaginary enclosure adjacent to the hood (Figure 6.19). The imaginary enclosure is not formed by walls or baffles but by air currents with the needed capture velocity moving into the hood. Any contaminant released inside this enclosure is captured; contaminants released outside will probably escape into the room atmosphere. Therefore, to design capturing hoods, you must determine the capture velocity and the location of contaminant release before calculating the hood's overall airflow requirements.

Figure 6.19   The capture zone around an open duct is a sphere bounded by velocity contours with the needed capture velocity. Contaminants released inside the sphere will be captured while those released outside may escape.

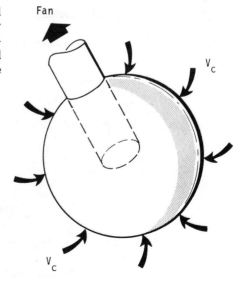

Capture Velocity

Recommendations for capture velocities for different operations are published in the ACGIH *Manual* (Table 6.5). These recommendations have survived the test of time and in general are adequate for most substances. However, very highly toxic materials may require a substantially higher velocity to capture enough material to prevent excessive airborne levels. Other factors that influence capture velocity guidelines are room air currents, the amount of time the hood will be used, and whether or not there is adequate general ventilation in the workroom. Be especially watchful for thermal air currents or other factors that disperse contaminants despite a capturing hood at the process.

Sometimes general capture velocity recommendations are not enough. For example, if you are selecting hoods for a new multihood ventilation system, it is important to have close estimates of the airflow needed at each hood. Otherwise the duct and fan design may be inaccurate. The best way to determine the needed capture velocity and airflow is to seek out similar equipment and operating conditions at other plants or to build a few hoods (even out of cardboard) and test their effectiveness at different airflows. Whatever the source, data showing degree of contaminant control as a function of airflow rate (see Figure 6.25) will help you select operating and design parameters for the system.

One last capture velocity consideration is the *null point* theory for capturing hoods developed by Hemeon (Figure 6.20).[1] Hemeon defines

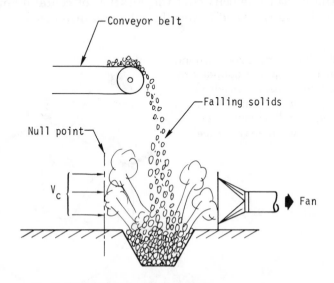

Figure 6.20 To control contaminants, the hood must generate the needed capture velocity at the *null point*, not just at the contaminant source.

Table 6.5 Range of Capture Velocities

Condition of Dispersion of Contaminant	Examples	Capture Velocity, ft/min
Released with practically no velocity into quiet air.	Evaporation from tanks; degreasing.	50–100
Released at low velocity into moderately still air.	Spray booths; intermittent container filling; low-speed conveyor transfers; welding; plating; pickling.	100–200
Active generation into zone of rapid air motion.	Spray painting in shallow booths; barrel filling; conveyor loading; crushers.	200–500
Released at high initial velocity into zone of very rapid air motion.	Grinding; abrasive blasting; tumbling.	500–2000

In each category above, a range of capture velocity is shown. The proper choice of values depends on several factors:

Lower End of Range	Upper End of Range
1. Room air currents minimal or favorable to capture.	1. Disturbing room air currents.
2. Contaminants of low toxicity or of nuisance value only.	2. Contaminants of high toxicity.
3. Intermittent, low production.	3. High production, heavy use.
4. Large hood—large air mass in motion.	4. Small hood—local control only.

Source: Reference 2.

the "null point" as the distance from the contaminant source where the integral air currents have expended their initial energy, and velocities have decreased to the magnitude of the random air currents. This theory states that the capture zone must cover not only the source of contaminants but must extend to the null point. Thus the capture velocity must be high enough to capture the contaminants at the location where initial air contaminant velocity dissipates. If calculation of airflow for the hood is based on developing the capture velocity only at the source, contaminants with a high enough initial velocity may escape from the hood capture zone. Typical applications for null point design include pulverizing applications, material handling if fines impact a hard surface and are dispersed, pneumatic tools if the exhaust disperses contaminants, and even grinding wheels that entrain air and generate high peripheral velocities.

Velocity Distribution

Proper distribution of the capture velocity is the second criterion for designing capturing hoods. The function of the hood is to create the "imaginary enclosure" (see Figure 6.19) that was described earlier. The "enclosure" is bounded by air moving toward the hood at the capture velocity. Contaminants released inside this zone are captured, while those released where the velocity is less than the needed capture velocity may escape into the workroom.

Equations describing how the velocity decreases with increasing distance from the hood have been available for 35 years. The problem is that many of these relationships refer only to the centerline velocity, that is, the air velocity along a line extending out from the center of the hood or duct. The equations describe the velocity at a point outside of the hood but do not define the velocity distribution across the hood face.[17] For example, the centerline velocity outside a free hanging plain hood is:

$$V_x = \frac{Q}{10X^2 + A} \tag{6.1}$$

where V_x = air velocity at X, ft/min
Q = airflow into hood, ft³/min
A = area of hood face, ft²
X = distance outward from hood along hood axis, ft

A plot of velocity as a function of distance (Figure 6.21) for a typical plain hood illustrates the limited reach of capturing hoods. But regardless of centerline velocity, the velocity contours in Figure 6.22 show that this hood type is inefficient for two reasons: it draws air from behind

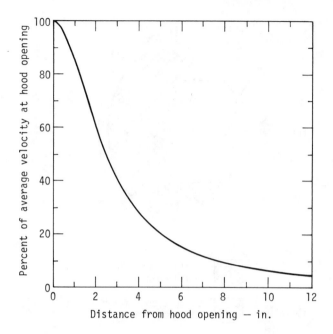

Figure 6.21 Percentage of hood face velocity as a function of distance from hood opening for a plain (unflanged) hood.

the hood outside the contamination zone; and the sharply bending airstreams flowing into the hood from behind interfere with smooth velocity contours in front of the hood where contaminants are generated. In other words, the velocity distribution is poor.

The plain hood can be easily improved by adding a flange to reduce the air drawn from behind the hood. This decreases the airflow requirement needed to develop the same V_x by about 25% for a flanged hood compared to a plain hood and changes Equation 6.1 to:

$$V_x = \frac{Q}{0.75\ (10X^2 + A)} \tag{6.2}$$

But more important than increasing centerline velocity, a flange improves the velocity distribution in front of the hood (Figure 6.22) and increases the size of the imaginary enclosure bounded by specific capture velocity contours. Similar velocity equations and velocity contours are available for other hood types (Figure 6.23).

In addition to flanged hoods there are three major ways to improve air distribution in front of capturing hoods:

- *Taper*—A gradual taper (Figure 6.24a) permits the inward flowing air to be drawn from the entire hood face. However, with too wide a

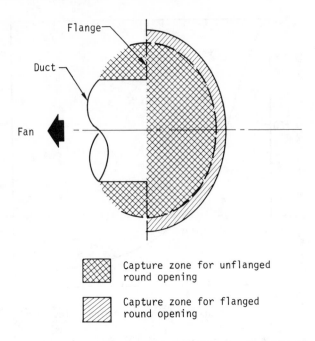

Capture zone for unflanged round opening

Capture zone for flanged round opening

Figure 6.22   A flange increases the size of the capture zone in front of a duct opening.

taper angle the airstreams break away from the taper and air distribution is not much better than with an included taper angle of about 60°. For large hoods, the physical size of the tapered section makes this velocity distribution technique impractical.

- *Splitter vanes*—The airflow into the hood can be channeled to different parts of the hood opening by using vanes inside the hood (Figure 6.24b). These have the advantage of causing fairly low resistance to airflow while saving space when compared to the gradual taper method of distributing airflow. However, in ventilation systems subject to corrosion, erosion, or material build-up, splitter vanes are vulnerable to attack from all of these sources.

- *Slot resistance*—A slot in a plenum chamber under suction is the third way to distribute airflow. Studies have shown that if the width (or height) of the slot does not exceed 20% of its length, air velocity along the slot will be fairly uniform (Figure 6.24c). Slots are used where other air distribution methods do not work. A good example is a long, open surface tank that has limited space for a hood yet has a need for good air distribution over its entire length. However, the term for this velocity distribution technique, *slot resistance*, should warn you of its main disadvantage. The slot is a narrow opening that creates high turbulent pressure losses as the air is acceler-

Hood type	Velocity contour shape	Airflow volume — centerline velocity equation
Plain opening	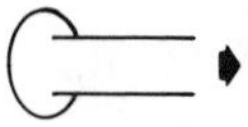	$Q = V_x(10X^2 + A)$
Flanged opening	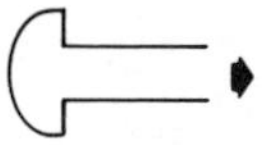	$Q = 0.75\ V_x(10X^2 + A)$
Slot	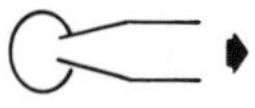	$Q = 3.7\ LV_x X$
Flanged slot	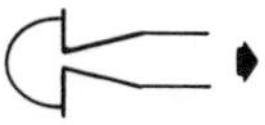	$Q = 2.8\ LV_x X$

Q = Airflow volume, ft^3/min
V_x = Centerline velocity at X, ft/min
X = Distance out from hood opening, ft
A = Area of hood opening, ft^2
L = Length of slot hood, ft

Figure 6.23   Velocity contour shape and volumetric airflow equations for different types of capturing hoods.

ated to the slot velocity and then decelerated inside the plenum and duct. A slot hood causes more resistance than other velocity distribution techniques. Extra energy needed to overcome the increased resistance must be added to the system by a larger fan than would otherwise be needed.

Also, do not make the mistake of thinking that a very high slot velocity significantly increases the reach of the hood. All the slot does is distribute the inward velocity along the length of the slot. A NIOSH-sponsored study of open surface tank ventilation showed that the capture velocity two feet out from a four-foot-long slot did not vary as the slot velocity ranged from 700 to 6000 ft/min. Since the hood entry loss varies with the square of the slot velocity, losses can be substantially reduced by selecting the lowest practical slot velocity. As a rule-of-thumb, a slot velocity of 2000 ft/min and a plenum velocity of 1000 ft/min give good air distribution without excessive pressure loss.

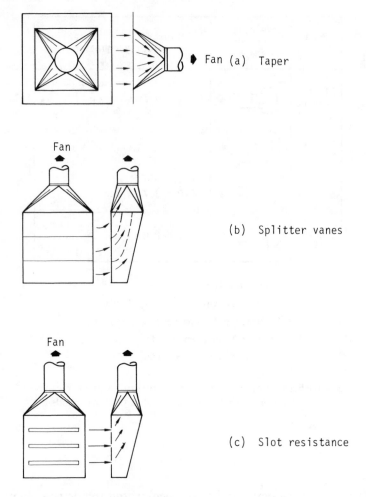

Figure 6.24   Three ways to distribute airflow into hoods. (Source: Reference 2)

The choice of the best method for distributing air velocity across the hood face depends on the space available for the hood and duct work and the size of the contaminant release zone. The goal is to achieve adequate distribution with minimum airflow volume while causing the least practical pressure loss through the hood.

The Problem of Limited Control

Capture velocity is a major factor in determining whether or not a capturing hood works properly. If you increase the capture velocity by increasing the airflow, the capturing hood should be more efficient.

Unfortunately, experiments show that this is not always the case. Past a certain capture velocity (depending on the hood type and size) additional increases in capture velocity result in little or perhaps no improvement in capture efficiency. At that capture velocity the hood is controlling about all it will ever control; random air currents or other factors allow some of the contaminants to escape.

For welding hoods, Figure 6.25 shows a typical fume concentration versus airflow relationship.[3] The level of iron oxide fumes in the welder's breathing zone decreases rapidly as the capture velocity increases from 0–600 ft/min. Essentially complete control is achievable at sufficiently high airflow rates.

Grinding wheel hoods represent a different case. With this type of hood the ventilation performance curve (Figure 6.26) shows that complete contaminant control is not attainable even at very high flow rates.[4] As the flow rate is increased to some point (about 250 ft³/min in this case), there is a steep descent in contaminant levels; any further increase has a minimal effect on concentration. According to a NIOSH study, this two-segment curve is typical for several grinding, buffing, and polishing machine ventilation hoods. The phenomenon of limited control with these

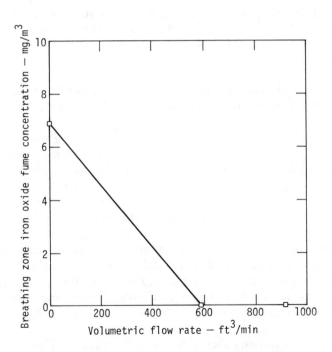

Figure 6.25 Breathing zone fume concentration as a function of airflow into a welding hood. Complete fume control is possible at high enough airflow rates.

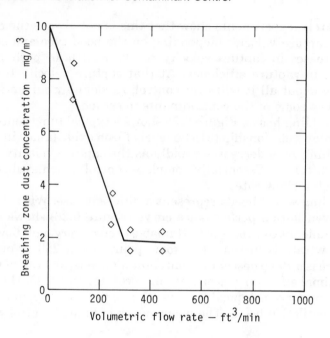

Figure 6.26   Breathing zone dust concentration as a function of airflow for a grinding wheel hood. Complete control is not possible even at high flow rates.

hoods can be explained if you realize that different size particles are generated by the spinning grinding wheel. The larger particles fly out tangentially and are easy to catch in a properly located receiving hood. Even if they are not caught, most will settle out (Figure 6.27a). Intermediate and fine particles remain close to the grinding wheel in the layer of air entrained by the spinning wheel. Intermediate size particles can be captured by a capturing hood if the capture velocity is high enough to overcome their curving path (Figure 6.27b). The very fine particles, however, are not captured but continue to move with the air layer around the wheel and are deflected into the workroom air by the workpiece being ground. Increasing the capture velocity controls more and more of the intermediate particles but has little effect on the very fine particles that stay near the wheel. Once the capture velocity is high enough to capture all the intermediate size particles, further increases will not reduce airborne levels since the airborne levels are due to the very fine particles.

To improve capture efficiencies, auxiliary hoods (Figure 6.28) were added to standard grinding hoods in the NIOSH-sponsored study.[4] The auxiliary hood was located to capture the very fine particles as they collided with the workpiece. Preliminary tests showed that an additional

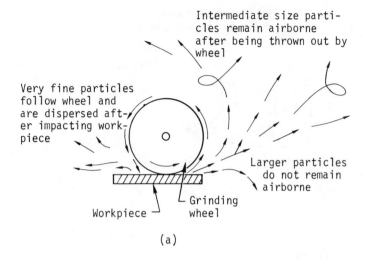

Intermediate size parti-
cles remain airborne
after being thrown out by
wheel

Very fine particles
follow wheel and
are dispersed aft-
er impacting work-
piece

Larger particles
do not remain
airborne

Grinding
wheel

Workpiece

(a)

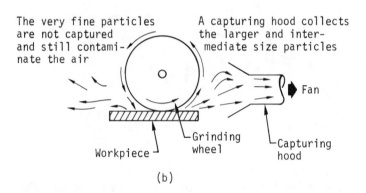

The very fine particles
are not captured
and still contami-
nate the air

A capturing hood collects
the larger and inter-
mediate size particles

Fan

Grinding
wheel

Workpiece

Capturing
hood

(b)

Figure 6.27 Complete control is not achieved in Figure 6.26 because the hood does not capture very fine particles traveling with the grinding wheel.

30–70% reduction in airborne contaminant levels could be achieved by using an auxiliary hood. Low-volume, high-velocity hoods (Figure 6.17) are another method to improve control efficiencies.

Slot hoods also have limits on their effectiveness, since the reach of the inward airflow does not increase indefinitely with increases in airflow. Remember that slot velocity has little effect on velocity in front of the slot; it only provides uniform velocity distribution over the length of the slot. For most applications a slot hood should not be designed to pull in contaminants more than two feet away. Thus tanks up to four feet wide can be exhausted using slots along both sides of the tank. These recommendations follow OSHA standards. However, recent

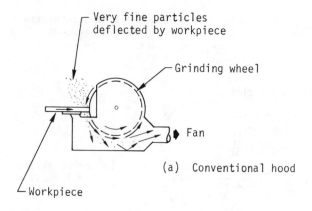

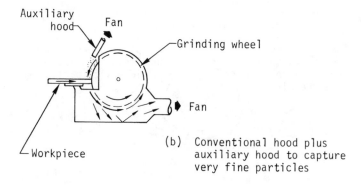

Figure 6.28   Adding an auxiliary hood to a conventional grinder hood reduces the quantity of very fine particles escaping into the workroom.

studies indicate that a single slot hood can reach out four or even eight feet and achieve partial control although the airflow rates are very high.[5] More information on choosing airflow rates for slot hoods is detailed in Chapter 3 on OSHA ventilation standards.

Push-pull slot hoods are useful on wide tanks. They increase collection efficiency and also reduce the exhaust airflow. A compressed air source on one side of a tank blows ("pushes") air across the tank toward a slot exhaust ("pull") on the other side of the tank. This action results in an air curtain that prevents contaminant release. Traditionally, push-pull systems (see Figure 6.18) have not been too effective because of poor air supply design and poor balancing between the supply air rate and the exhaust rate.

If the supply airflow is too great compared with the exhaust rate, the supply air is deflected into the workroom. If the supply rate is too low, then the exhaust slot works as a pull-only system and does not get the benefit of the air supply. There is an optimum balance (Figure 6.29) for tanks from four to eight feet wide.[5] For example, if the exhaust rate is 100 ft³/min of air for each square foot of tank surface, the optimum air supply rate is about 16 ft³/min for each foot of tank length. Other values can be read from the graph for exhaust rates ranging from 75–150 ft³/min per square foot of tank surface. However, beyond about 100 ft³/min per square foot of tank surface the cost of the exhaust air system begins to mount rapidly. Even when the air supply and exhaust rates are properly balanced, push-pull systems have these inherent disadvantages:

- Large objects disrupt the air curtain when they are lowered into the tank or removed from it. Controls, automatic or manual, can be used to stop the air supply during these operations.
- The air curtain increases turbulence at the liquid surface, thereby increasing evaporation of tank contents and heat loss from hot baths. For some solvents the evaporation loss is an economic problem.

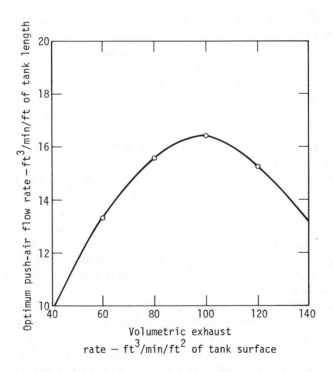

Figure 6.29 Optimum push airflow rates for push-pull systems as a function of the volumetric exhaust rate.

Detailed design recommendations for push-pull and other hoods are contained in the ACGIH *Manual.*

Crossdrafts Disrupt Capturing Hoods

The imaginary enclosure (see Figure 6.19) formed by velocity contours outside a capturing hood is fragile. Nothing is more disruptive than crossdrafts in the room or other random air currents from open windows, fans, or space heaters. Design standards from the ACGIH *Manual* or other sources no longer apply when significant drafts exist in the work-room. For example, open surface tank emissions double for crossdrafts of 60 ft/min compared with a location having no appreciable drafts (see Figure 6.15). Unfortunately there is no easy way to consider these effects during hood selection and design. If the work location has significant crossdrafts, they must be eliminated, deflected with baffles or curtains, or dealt with by using another hood type rather than a capturing hood.

Enclosures

After all the pitfalls and design tips for capturing hoods, designing en-closures will seem easy. For capturing hoods, the large airflow necessary for extending the capture zone to regions outside the hood is a decided disadvantage. Enclosures avoid this by enclosing the source with walls and doors with minimum open area. Most enclosures are designed under the assumption that workers do not enter (or at least insert their heads into) the enclosure during the operation without suitable respiratory protection.[7] Thus the levels of airborne contaminants inside the en-closure are not important as long as flammable or explosive mixtures do not form. By definition, an enclosure is a hood that surrounds the point of emission so that all initial dispersive action of the contaminant takes place within the hood. Why qualify the definition by requiring that the contaminants' dispersive action occur within the hood? The reason is that enclosures are designed to create sufficient inward air velocity through openings to contain material released inside the hood. In this way the material does not escape from the enclosure; few enclosing hoods have high enough face velocity to overcome measurable outward velocity due to dispersive forces.

For example, suppose you are designing an enclosing hood for a quenching operation. Small billets of lead are heated in an oven, pressed into shape, and cooled by immersing them in a 55-gallon drum containing water (Figure 6.30a). A drum filled to within two feet of the top is large enough for the quenching job. The problem is that the steam cloud formed during quenching quickly rises three or four feet and escapes from the drum. An enclosure for the drum will have to extend the drum height

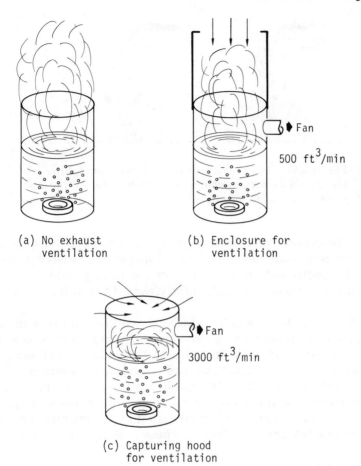

(a) No exhaust
    ventilation

(b) Enclosure for
    ventilation

(c) Capturing hood
    for ventilation

Figure 6.30 Exhausting a 55-gallon drum used to quench hot lead pressings. With no ventilation (a) lead fumes escape with the steam cloud. An enclosure large enough to surround the steam cloud (b) requires 500 ft³/min to control the fumes. With no enclosure (c) 3000 ft³/min is needed to stop and capture the steam and fumes.

to contain this cloud (Figure 6.30b). Otherwise the hood will not be an enclosure but a capturing hood (Figure 6.30c) that requires about 600% more airflow than a properly designed enclosure to overcome the upward velocity of the steam cloud.

Another good illustration is a falling stream of fine solids like sawdust or fertilizer. The very fine particles disperse into a cloud when they strike a hard surface. The size of the cloud is increased because the falling material entrains air into the stream of solids (Figure 6.31). An

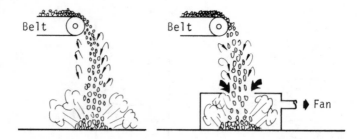

Figure 6.31   An enclosure to control dust from falling solids should be large enough to surround the initial dust cloud and the pile of solids.

enclosure to contain the cloud of very fine particles must be sized to contain the dust cloud after the initial dispersive forces are spent. To remove the additional air entrained in the falling solids, the exhaust rate from the enclosure should slightly exceed the ordinary inward air velocity.

Although enclosures help reduce exhaust air requirements for any process, they are practically a necessity for some operations involving very highly toxic materials. Included in this category are many radioactive materials, some human pathogens, and some carcinogens (cancer-causing agents) for which no detectable worker exposure is allowed. For these compounds complete enclosures such as glove boxes (Figure 6.32) can reduce routine exposures to near zero if proper operating procedures and safety precautions are followed. These systems are covered in Chapter 10.

Face Velocity is the Key

The important design parameter for enclosures is the inward air velocity through cracks, doors, and other openings. When the velocity is measured in the plane of the opening, it is called *face velocity*. The face velocity must be high enough to keep contaminants from escaping despite random room air currents, workers walking past the openings, and other disruptive forces. Often velocity criteria are stated as an average inward face velocity with an additional requirement that the velocity at all points be at least 75 or 80% of the average face velocity. For example, a laboratory fume hood (see Figure 6.11) might be purchased to develop 100 ft/min average face velocity with a minimum at any point in the hood opening of 80 ft/min. Face velocity recommendations for all hoods depend on the toxicity of materials handled inside the hood, the amount of time the hood is used, and the operations controlled by the hood. Typical recommendations are listed in Table 6.6.

Figure 6.32 Glove boxes isolate highly toxic materials from workers and the outside environment. *(Courtesy Kewaunee Scientific Equipment Corp.)*

Table 6.6 Face Velocity Guidelines for Enclosures[a]

Operation	*Face Velocity,*[b] *ft/min*
Welding booth	150
Paint spray booth	100–200
Laboratory hood	100–150
Abrasive blasting room	500
Belt conveyor enclosures	150–200
Bin or hopper enclosure	150–200
Mixer	100–200
Metallizing enclosure	
toxic	200 (with respirator)
nontoxic	125
Melting furnace	150–250 + products of combustion, if any
Swing frame grinder booth	150
Machining toxic materials	300

[a]See OSHA standards for specific chemicals or operation to see whether legal standards apply.
[b]Face velocity is also expressed as ft³/min of air per ft² of openings into hood or enclosure. Units cancel to yield velocity units (ft/min).

Once the proper face velocity has been selected, the airflow can be calculated using this equation:

$$Q = V_{av} \times A_{opening} \times F_s \qquad (6.3)$$

where Q = airflow volume, ft³/min
V_{av} = average face velocity, ft/min
$A_{opening}$ = area of openings in enclosure, ft²
F_s = dimensionless safety factor to account for poor air distribution through the enclosure openings, typically 1.0–1.5

The magnitude of F_s depends on two factors: the amount of excess air that is needed and the minimum amount of air needed to achieve the face velocity that yields the minimum-velocity-at-any-point criterion. If the inward flowing air is distributed fairly uniformly across the hood openings, the safety factor value approaches 1.0.

To minimize the airflow volume requirements, air distribution baffles are usually installed in open face booths and other enclosures with large open areas. Without some method of air distribution, face velocity into these booths is highest toward the center and drops off at the edges (Figure 6.33). The baffles generally occupy 40–75% of the booth face area and are located near the back of the booth (Figure 6.34). If the exhaust duct is connected through the rear wall, the baffles are usually vertical; for top duct connections the baffles are angled. For small face areas a single baffle plate is sufficient to distribute the air properly; for booths with open areas exceeding about 3 ft × 3 ft multiple baffles are used. A plenum chamber on the back of the enclosure with slots to draw

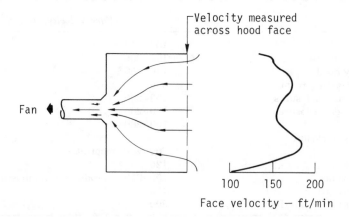

Figure 6.33 Velocity distribution across the face of an open booth varies and is usually lowest at the edges.

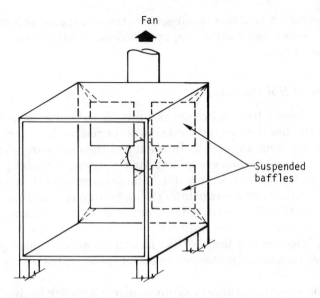

Figure 6.34 Suspended baffles distribute the air-flow to equalize velocity across the face opening.

air across the openings (see Figure 6.12) is another way to distribute air across an enclosure with large open areas.

Many enclosures have openings equipped with sashes that can be closed during some operations. Laboratory hoods with sliding sashes are typical; the tempered glass in the window acts as a safety shield when the sash is lowered. The sash can be opened fully for setup, then closed partially or fully, depending on hood design, for periods when contaminants are released inside the hood. Some airflow is needed for the fan to exhaust contaminants properly. If closing the sash cuts off all this inward airflow, the sash should be kept at least partially open. As the sash is closed, the inward face velocity increases for most hoods. The question is whether the system should be designed to develop the required velocity with doors fully opened or partially closed? The answer is important if you are faced with upgrading substandard hoods, and it depends on these factors:

- OSHA standards covering your location. Federal OSHA standards do not address this specifically although some state standards may.
- Type of operation. If contaminants can be released with enough energy to escape from the hood despite the inward face velocity, the doors should be closed during the operation. In these cases the airflow can be calculated assuming the doors are partially closed.

- Administrative controls, such as written operating procedures plus adequate supervisory attention, are needed so that workers keep doors positioned properly.

Beware of Hot Processes

When an enclosure houses hot operations, there is one more design consideration: the heated air and contaminants rise with a "thermal head" and will escape from any cracks in the top of the enclosure (Figure 6.35). You can visualize the process as hot air rising rapidly and piling up in the upper part of the enclosure until the fan removes it. Any openings allow the contaminants to continue rising right out of the enclosure into the workroom.[1] The solution is to make the construction as airtight as possible. If openings are required near the top, make them as small as possible. Design the fan so that it will keep the upper part of the enclosure at negative pressure, even at peak rates of contaminant generation.

For additional discussion of enclosure hoods for highly toxic materials, see Chapter 10.

Receiving Hoods

Receiving hoods are the third major type of hood. They are positioned to catch a stream of contaminants or contaminated air thrown out in a

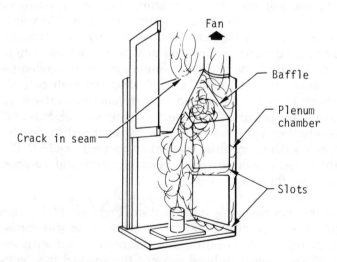

Figure 6.35 Heated contaminated air can escape through openings at the top of an enclosure if the fan is too small to keep the top at negative pressure.

given direction by an industrial process. Receiving hoods may look like capturing hoods but they are different. Capturing hoods have a higher airflow rate; they can reach outside the hood to draw in contaminants. Receiving hoods sit and wait for contaminants to enter the hood; the exhaust flow rate is calculated from the volume of air entering the hood.

Receiving hoods are covered last in this chapter because they have limited value in health protection ventilation systems. Unless they are carried by a stream of air, the very fine particulates along with hazardous gases and vapors presenting health hazards do not travel far enough to reach a receiving hood. These small contaminants meet so much resistance in traveling through air that they have no power of independent motion apart from the air surrounding the contaminants. However, receiving hoods do a good job controlling large particles such as sawdust and chips of nontoxic materials from machining operations.

The best places to use receiving hoods are where the layout of the process makes an enclosure impractical and the contaminants have enough energy or initial motion to travel into the hood. Popular applications include:

- Grinding, sawing, sanding, and polishing operations using low-toxicity materials (Figure 6.36). The hood should be shaped to fit as closely as possible around the moving machinery, and openings into the hood located to catch particles that are thrown off. Since the very fine dust will escape into the workroom, the materials being worked and the materials in the abrasives must be nontoxic.
- Canopy hoods over hot processes. Since heated air and vapors rise, the easiest control method may be a canopy hood over the process (Figure 6.9). The canopy cannot be used when workers must lean over

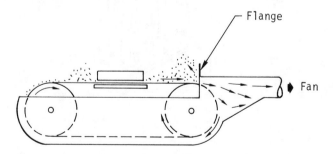

Belt grinder or polisher

Figure 6.36 Receiving hoods, such as this belt grinder or polisher hood, control large particles. Although small particles may escape, receiving hoods are often acceptable for nuisance or low-toxicity materials.

the tank or process; as contaminants rise, workers will breathe the contaminated air.

Receiving hoods are sized to prevent overloading and random spillage. These factors are especially important for canopies over hot processes. Overloading (Figure 6.37) occurs when more contaminated air enters the hood than is removed by the exhaust fan.[1] The excess overflows into the room. The exhaust flow rate is calculated from estimates of the peak flow rate into the hood during any part of the process, including room air entrained into the rising contaminated air:

$$Q_{total} = Q_{contaminant} + Q_{entrained} \qquad (6.4)$$

where Q_{total} = design exhaust flow rate, ft³/min
$Q_{contaminant}$ = estimated contaminant or contaminated airflow rate from process, ft³/min

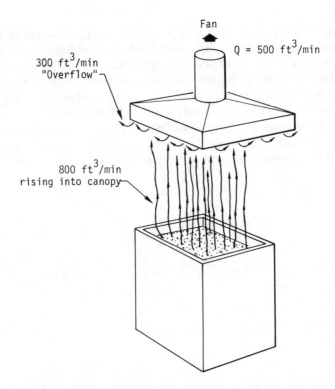

Figure 6.37 "Overflow" of contaminants from a receiving hood occurs if more contaminated air enters the hood than the fan exhausts.

$Q_{entrained}$ = estimated amount of room air entrained into contaminant stream, ft³/min

For canopies over hot processes a deep skirt around the hood bottom can reduce overloading, especially if the volume of emissions from the process varies sharply over short time spans.

Random spillage occurs when crossdrafts or other random air currents cause contaminated air to escape from the hood. Spillage is especially noticeable in canopies that are significantly larger than the cross-sectional area of the rising column of contaminants. This is because there is a dead air space between the rising contaminants and the hood edge (Figure 6.38). The way to avoid random spillage is to increase the exhaust flow, thereby removing the air rising into the hood as well as developing an inward velocity around the perimeter of the hood. An inward velocity of 100–150 ft/min is typical, depending on the magnitude of the crossdrafts in the room and the toxicity of the contaminants. Sides, curtains, and baffles can also reduce random spillage. When random spillage is a problem, exhaust rates can be estimated from:[1]

$$Q_{total} = Q_{contaminant} + (V_{edge} \times P) \qquad (6.5)$$

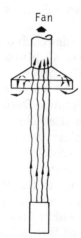

(a) If rising column does not fill hood, random spillage can occur

(b) Cross-sectional view showing turbulence inside canopy

Figure 6.38 Random spillage occurs when air currents and drafts blow air into and out of a canopy hood. It is more noticeable when the hood opening is larger than the contaminant stream. (Source: Reference 1)

where Q_{total} = design exhaust flow rate, ft³/min
  $Q_{contaminant}$ = estimated contaminant or contaminated airflow rate from process, ft³/min
  V_{edge} = inward velocity at hood edge, ft/min
  P = perimeter of hood edge, ft

No Canopies for Cold Processes

Canopies for unheated processes are not designed as receiving hoods unless the contaminants are lighter-than-air or are projected upward into the hood by some other force. Canopies for unheated processes must be designed as capturing hoods, although the tremendous airflows needed to develop capture velocities two or three feet below the canopy usually make these hoods impractical. The answer is to select another type of capturing hood, such as a side draft or slot hood, or an enclosure.

SUMMARY

There are three different types of hoods used in local ventilation systems: capturing hoods, enclosures, and receiving hoods. Each works according to a different principle to control contaminants:

- Capturing hoods reach out to capture contaminants in the workroom near the hood. Capture velocities range from 50–2000 ft/min depending on the process, the toxicity of materials released, and the conditions in the workroom. Since air velocities drop off sharply with increasing distance from the hood opening, it is most economical to locate the hood as close to the contaminant source as possible and rarely more than two feet away.
- Enclosures surround the contaminant source and are designed to keep the contaminants from escaping. The exhaust airflow is based on the area of openings in the enclosure and the inward air velocity through the openings needed to prevent contaminant escape. Due to low exhaust rates, enclosures are usually the most economical hoods to install. Inward face velocities of 100–150 ft/min are typical.
- Receiving hoods are positioned to catch contaminants emitted from a source along a specific path. Exhaust rates are based on the volume of contaminants or contaminated air entering the hood. Receiving hoods have limited application in health protection ventilation systems except for canopy hoods over heated processes. This is because they do not always control fine particulate or gaseous or vapor contaminants. These types of contaminants pose more of an inhalation hazard than the large particles that receiving hoods do control.

Selecting the proper hood depends on the contaminant. Gases, vapors and the very small particulates in the respirable size range have no independent motion of their own; they move with the air around them. These contaminants are controlled by controlling the air they move with. Particulates with aerodynamic sizes exceeding about 50 μm are too large to remain airborne and so rarely present an inhalation hazard. Hence controlling them is not usually important from a health standpoint.

The important factor in hood selection is that each hood type works on a different principle. Confusing or violating the principles means that the hood will not function properly in controlling contaminants.

REFERENCES

1. Hemeon, W.C.L. *Plant and Process Ventilation* (New York, New York: Industrial Press, Inc., 1963).
2. ACGIH Committee on Industrial Ventilation. *Industrial Ventilation—A Manual of Recommended Practice,* 17th Ed. (Lansing, Michigan: American Conference of Governmental Industrial Hygienists, 1982).
3. Astleford, W. "Engineering Controls of Welding Fumes," U.S. Department of Health, Education and Welfare, Publication No. (NIOSH) 75–115 (Washington, D.C.: U.S. Government Printing Office, 1974).
4. Bastress, E.K., J.M. Niedzwecki, and A.E. Nugent. "Ventilation Requirements of Grinding, Buffing and Polishing Operations," U.S. Department of Health, Education and Welfare, Publication No. (NIOSH) 75–107 (Washington, D.C.: U.S. Government Printing Office, 1975).
5. Flanigan, L.J., S.G. Talbert, D.E. Semones, and B.C. Kim. "Development of Design Criteria for Exhaust Systems for Open Surface Tanks," U.S. Department of Health, Education and Welfare, Publication No. (NIOSH) 75–108 (Washington, D.C.: U.S. Government Printing Office, 1975).
6. Lynch, J.R. "Industrial Ventilation: A New Look at an Old Problem," *Michigan's Occupational Health* 19, No. 3 (1974).
7. American National Standard Z 9.2–1971. "Fundamentals Governing the Design and Operation of Local Exhaust Systems," (New York, New York: American National Standards Institute, 1972).
8. Olishifski, J.B. and F.E. McElroy, Eds. *Fundamentals of Industrial Hygiene* (Chicago, Illinois: National Safety Council, 1971).
9. Sax, N.I. *Dangerous Properties of Industrial Materials* (New York, New York: Van Nostrand Reinhold Company, 1975).
10. OSHA General Industry Safety and Health Regulations, U.S. code of Federal Regulations, Title 29, Chapter XVII, Part 1910.1006 (1975).
11. National Institute for Occupational Safety and Health. *The Industrial Environment—Its Evaluation and Control* (Washington, D.C.: U.S. Government Printing Office, 1973).
12. Goodier, J.L., E. Boudreau, G. Coletta, and R. Lucas. "Industrial Health and Safety Criteria for Abrasive Blast Cleaning Operations," U.S. De-

partment of Health, Education and Welfare, Publication No. (NIOSH) 75–112 (Washington, D.C.: U.S. Government Printing Office, 1975).

13. Davison, R.L., D.F.S. Natusch, J.R. Wallace, and C.A. Evans. "Trace Elements in Fly Ash, Dependence of Concentration on Particle Size," *Environmental Science and Technology* 8, No. 13, 1107 (1974).

14. American Medical Association. "The Wonderful Human Machine," (Chicago, Illinois: American Medical Association, 1961).

15. American Conference of Governmental Industrial Hygienists. "TLVs—Threshold Limit Values for Chemical Substances in the Workroom Environment," (Cincinnati, Ohio: ACGIH, 1983).

16. Staheli, A.H. "Control Methods for Reducing Chlorinated Hydrocarbon Emissions from Vapor Degreasers," ASME Reprint No. 72–PEM–7 (New York, New York: The American Society of Mechanical Engineers, 1972).

17. Dalla Valle, J.M. *Exhaust Hoods* (New York, New York: The Industrial Press, Inc., 1952).

18. Caplan, K.J. and G.W. Knutson. "A Performance Test for Laboratory Fume Hoods," *Amer. Industrial Hygiene Assoc. J.* 43, No. 10, 722 (1982).

19. Peck, R.C. "Validation of a Method to Determine a Protection Factor for Laboratory Hoods," *Amer. Industrial Hygiene Assoc. J.* 43, No. 8, 596 (1982).

20. Caplan, K.J. and G.W. Knutson. "Influence of Room Air Supply on Laboratory Hoods," *Amer. Industrial Hygiene Assoc. J.* 43, No. 10, 738 (1982).

CHAPTER 7

Air Cleaner Selection

This chapter covers air cleaner selection. The term *selection* is more appropriate than the term *design* because air cleaners are packaged units that are purchased and installed in local exhaust systems. Few system designers ever design an air-cleaning device and have it fabricated.

This discussion of air cleaners focuses on local exhaust systems that collect contaminants in the workplace. The gas to be cleaned consists of room air and the captured contaminants. Air cleaners for boilers or other combustion processes are not covered.

Although the air cleaner is an add-on device in many exhaust systems, it often represents the greatest single source of resistance in the system. If a unit that causes relatively low pressure drop can be used, fan power costs over the life of the system will be reduced. This, of course, must be balanced against capital cost of the air cleaner, and operating and maintenance costs.

Exhaust systems that handle high toxicity or high nuisance contaminants may require special air cleaners, operating controls, and fan arrangements to assure a degree of emissions control and operating reliability not needed in typical industrial exhaust systems. Chapter 10 covers design and operating features, including air cleaners, for these systems.

IS AN AIR CLEANER NEEDED?

The first step in air cleaner selection is deciding whether an air cleaner is needed at all. The answer depends on these four factors:

- The toxicity of the material carried in the system and whether discharging it into the outdoor environment could create potential hazards to employees or the surrounding community.

- The amount of material carried in the system and whether discharging it to the environment could create a housekeeping or esthetic problem. For example, an air cleaner is required to remove the sawdust and wood chips collected in the local ventilation system of a woodworking shop. Otherwise the wind-blown pile of sawdust outside the shop would soon bury the building.
- The value of material carried in the ventilation system and whether it has significant salvage or recycle value. If its salvage value exceeds the cost of the collector, then it pays to include the air cleaner in the system. For example, solvent recovery systems are becoming popular on solvent vapor degreasers used for metal cleaning. Traditionally the large amount of solvent lost from the degreaser into a ventilation system made exhausting these units difficult to justify. However, OSHA regulations have focused attention on the employee exposure to solvent vapors escaping from the degreasers. Although ventilation is only one of several methods for controlling vapor emissions, it is sometimes the most practical solution. For these installations a vapor recovery unit can reduce solvent losses by 85–90%.[1]
- In many locations government regulations are a major factor in deciding whether an air cleaner is needed. The U.S. Environmental Protection Agency regulates the amount of hazardous air pollutants such as mercury, beryllium, asbestos, vinyl chloride, benzene, and radionuclides that can be discharged from certain industrial operations.[2] Although air cleaners *per se* are not specified, they are required in many cases in order to meet the standards. Federal OSHA standards require air-cleaning devices on ventilation systems used for certain carcinogenic substances. In some areas local air pollution regulations require a permit before installing almost any ventilation system that can discharge pollutants into the environment. In many other jurisdictions there are no specific air regulations pertaining to ventilation system discharges as long as they do not create a nuisance or endanger public health.

EXHAUST GAS CHARACTERISTICS

In some cases a decision on whether an air cleaner is needed cannot be made until the contaminant levels and other exhaust gas characteristics have been determined. If this analysis shows that the system requires a cleaning device, the exhaust characteristics can be provided to the equipment vendor as part of the specification package. Analysis may also help complete the application for an air pollution permit, if one is required.

The following information about the exhaust gas should be determined:

- Identity of contaminants and their physical state.
- Quantity of each contaminant released, both the average value and the short-term peak rate during batch dumping or similar operations that cause a high rate of release.
- For particulates, the size distribution and any shape, density, or other characteristics that may interfere with collection in some devices, for example, sticky droplets.
- Exhaust gas volumetric flowrate—average, short-term peak, and minimum values.
- Exhaust gas temperature and humidity, since both influence the type of collector that may be selected. High temperature also increases the volume of exhaust gas to be treated. High humidity can cause a visible steam plume at the stack outlet; if this is objectionable the exhaust may have to be reheated prior to release.

This information can be used to estimate the possible nuisance or health hazard from the exhaust discharge as well as the degree of compliance with air pollution regulations. The identity and concentration of the contaminants will allow a vendor to evaluate whether devices based on solubility or chemical reactivity will be effective. For particulates, the required collection efficiency for different size particles is the most important parameter, since expense rises rapidly with the need to remove small particles.

IDENTIFYING CONTAMINANTS AND ESTIMATING LEVELS

The task of identifying the exhaust gas contaminants and their concentrations is approached differently depending on the type of system that is involved (Table 7.1).

Table 7.1   Characterizing Contaminants for Air Cleaner Selection

Type of System	Source of Information
New Air Cleaner for Existing System	Air pollution stack sampling.
New Exhaust System for Existing Process	Air sampling at each proposed hood location.
	Published studies.
New Exhaust System for New Process	Air sampling at similar equipment.
	Vendors'/consultants' experience.

Existing Ventilation System

If the air cleaner is to be installed as a modification to an existing local exhaust system, traditional stack sampling techniques used in air pollution studies will provide the necessary information.[3] Sampling should be performed not only in the exhaust stack, but in each of the main ducts as well so that each of the individual exhaust streams may be analyzed. For particulate contaminants it is especially important to follow proper stack sampling procedures. In isokinetic sampling, for example, the sampling rate is adjusted at each sampling point in a duct so that the velocity of air drawn into the sampling probe is equal to the duct velocity. The result is samples that accurately reflect the mass concentration and the particle size distribution in the exhaust gas.

New Ventilation System
For Existing Operations

If a local exhaust system is being designed for an existing process that is not now connected to a local exhaust, the task of estimating contaminant levels is more difficult.

Air measurements at the location of each planned hood will usually give a good idea of the identity of contaminants, but often will not provide an estimate of the release rate. Temporary enclosures made of plastic sheeting or cardboard may be erected around release points so that all of the released contaminant can be contained for collection by a high-volume sampling probe. As described earlier, for particulates the collection and analytical methods should be selected to yield size distribution as well as mass concentration information.

Keep in mind that emissions from some equipment, such as open surface tanks, may be increased once the local exhaust system is installed. This is due to additional evaporation or droplet formation caused by air moving across the tank surface.

New Exhaust System For
New Operation

Where a new operation is to be exhausted, the task of estimating contaminant loading in a meaningful way can be extremely difficult. If similar equipment is in operation at the plant or even at another facility that is accessible (typically a different plant within the same company), air measurements as described above can give at least a rough guess of what contaminant loading can be anticipated.

Published studies may help when nothing better is available. For example, welding fume generation rates have been studied by the American Welding Society and other organizations under a wide variety of conditions.[4] Table 7.2 shows the range of fume generation rates in units of milligrams of fume per minute for different categories of welding rods.[5] This information can be used to estimate the overall level of fumes in the work environment and also in the duct of a local exhaust system.

Table 7.2 Welding Fume Generation Rates

AWS[a] Electrode Classification	*Type/Application*	*Fume Generation Rate (mg/min)*
E6010	Mild steel electrode—welding castings, pressure vessels, galvanized plate.	480–800
E6013	Mild steel electrode—welding thin/medium gauge steel: auto bodies, building structures.	300–600
E7018	Low hydrogen steel electrode (iron powder)—welding carbon and low-alloy steels.	600
E7024	Mild steel electrode—high speed, high deposition fillet and lap welds.	480–520
E316–15	Stainless steel coated electrode—fabricating 316 stainless steel equipment for chemical service.	300–400

[a]*American Welding Society.*
Source: Reference 5.

Example: Five welders are using E7018 shielded metal arc welding rods and are operating at about 50% arc time, that is, they are actually welding about 50% of the time. Estimate the concentration of welding fumes in a local exhaust system discharge stack if each welder has a close capture hood (see Figure 6.6) drawing 500 ft³/min of air.

Answer: From Table 7.2, each welder produces about 600 mg/min while welding.
Peak rate equals:

$$5 \text{ welders} \times 600 \, \frac{\text{mg/min}}{\text{welder}} = 3000 \text{ mg/min}$$

Airflow through the system equals:

$$5 \text{ welders} \times 500 \, \frac{\text{ft}^3/\text{min}}{\text{welding hood}} = 2500 \text{ ft}^3/\text{min}$$

Converting to metric units:

$$2500 \, \frac{\text{ft}^3}{\text{min}} \times \left(\frac{\text{M}^3}{39.3 \text{ ft}^3} \right) = 70.8 \text{ M}^3/\text{min}$$

Peak concentration in the exhaust stream while all arcs are in operation equals:

$$\frac{3000 \text{ mg/min}}{70.8 \text{ M}^3/\text{min}} = 42 \text{ mg/M}^3$$

Average concentration over the work day in the exhaust air is 50% of this value, or about 21 mg/M³.

Depending on the rods, the specific contaminants generated by the metal being welded and where the exhaust gas is discharged, this calculated fume concentration may indicate whether or not an air cleaner is needed for this system.

When determining the exhaust gas characteristics, any lack of information needed to make a sound estimate should be counterbalanced by providing additional capacity in the air cleaner or by designing a system that permits easy expansion by adding additional air cleaning units in series or in parallel.

AIR CLEANING DEVICES

The ideal air cleaner for a specific application would have these features:

• Low cost (initial and operating).
• High efficiency for the contaminants.
• No decline in operating efficiency with time, either between periodic cleaning cycles or over the useful life of the unit.
• Continuous operation during the work period without unscheduled shutdowns that interfere with production.
• Normal maintenance and disposal of collected material without hazardous employee exposures.

The types of devices to consider depend primarily on the physical state of the contaminants, whether they are particulates or gases/vapors. With the possible exception of combustion devices for organic fumes (extremely fine particles) and odorous vapors, there is no single device that is highly efficient for both small particulates and for gases/vapors. Scrubbing devices are widely used to collect some particles and gases or vapors in a single unit, but these combination units are not highly efficient for fine particles.

PARTICULATE REMOVAL

Air cleaners for particulates are listed in Table 7.3 along with the principle of operation and typical particle size for 90% collection efficiency. Not included in this discussion are simple inertial separators that reduce air velocity so that larger particles settle out. These separators are not effective for respirable size particles but may be useful as a primary collector in a two-stage collection system that also contains a more efficient unit.

Table 7.3   Particulate Air Cleaning Devices

Device	Cleaning Mechanism	Particle Diameter for 90% Removal (μm)
Baghouse	Filtration	>1
Electrostatic Precipitator	Electrostatic Attachment	>1
Cyclone—Small Diameter	Centrifugal Force	>5
Cyclone—Large Diameter	Centrifugal Force	25
Scrubber—Spray Chamber	Inertial Impingement	25
Scrubber—Packed Bed	Inertial Impingement	5
Scrubber—High Energy (Venturi, etc.)	Inertial Impingement and Centrifugal Force	>1

While particle size distribution is the most important factor in determining particulate removal efficiency, particle shape and density can also have significant effects. Nonspherical shapes and low densities tend to lower terminal velocities of the particles and hence the effectiveness of inertial collectors, which throw the particles out of the exhaust stream. The electrical properties of the particles are important if an electrostatic precipitator is to be used, since a particle must accept and hold a charge for the device to function properly.

These parameters may seem to rule out some collectors early in the evaluation process. However, it may be possible to use a water spray or other method to increase particle size by agglomeration or to change the electrical properties.

Filters

Filters trap particulates as the exhaust gas flows through a porous medium. Filters of woven or felted (pressed) fabric are most common, although paper and woven metal are also used. Filters have the general advantage of being able to handle varying exhaust gas flowrates and particle loadings.

Filters collect particles by three mechanisms:[6]

- *Impaction* occurs when particles have so much inertia that they collide with filter fibers rather than flowing around them.
- *Interception* occurs when particles do not actually hit the fibers but come so close that they are caught in the low-velocity air layer that surrounds each fiber.
- *Diffusion* results when small particles move randomly back and forth within the filter bed due to Brownian Motion (collisions with air molecules).

Impaction and interception are the key collection forces for particles larger than 1 μm; diffusion predominates below 0.1 μm. Between 0.1 and 1 μm all three mechanisms contribute to collection efficiency.[6] Since both impaction and interception are increased with small, closely spaced fibers, high efficiency filters exhibit high resistance to airflow.

Filter devices fall into two major categories:

- Disposable filters that often use inexpensive materials and are available in different configurations such as panels, mats, and cartridges. The High Efficiency Particulate Air (HEPA) filter is a special disposable filter used mainly for high toxicity particulates as described in Chapter 10.
- Reuseable filter elements in a housing that is equipped with a cleaning mechanism for periodic dust removal.

Selection of disposable or reusable filters is based on the expense of replacing the elements versus the added initial cost of the filter cleaning mechanism.

A common reusable filter unit for industrial exhaust systems is the baghouse (Figure 7.1). It consists of tubular fabric filters arranged in a housing along with the cleaning mechanism, which can be an automatic or manual shaking device, a means of blowing air back through the bags from the clean side, or a method of dislodging the accumulated dust cake. Chunks of the cake should be large enough so that they are not reentrained in the exhaust gas stream. Baghouses can collect practically all particles greater than 1 μm as well as a large percentage of submicron particles. Typical pressure drop is 2–8 inches of water.[6]

Prior to their initial use, new fabric filters have open spaces between fibers. Thus, the early collection efficiency is below the ultimate value achieved after several use/cleaning cycles when fine particles become embedded in the filter media. In addition, some filtering efficiency for small particles is temporarily lost after each cleaning cycle. This is because part of the overall filtering is due to the dust cake that accumulates on the fabric. This loss in efficiency for small particles can be important if all filter elements in a unit are cleaned at once. The temporary loss in efficiency in all filters at the same time may increase emissions enough to cause problems. Large baghouses have several compartments; if only one compartment at a time is cleaned, the overall stack emissions will be lower since the other compartments are still operating at high efficiency. Both filtering efficiency and pressure drop increase exponentially during cake repair. Then efficiency remains constant and pressure drop increases linearly until cleaning is needed.

Bag failure is probably the biggest problem with baghouse-type reusable filters. Generally, the problem manifests itself as a gradual

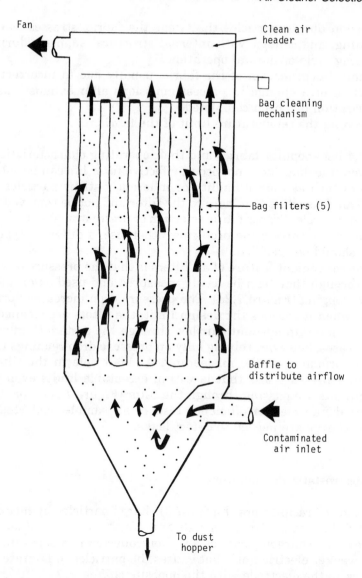

Figure 7.1 Cross-sectional view of a typical baghouse air cleaner. The bag cleaning mechanism, shown symbolically in this figure to illustrate a key feature of the unit, could be reversed air jets, a mechanical shaker, or another device.

increase in stack emissions rather than a sudden failure of one or more bags. In small units, bags are often replaced as a set because of the difficulty in locating the single leaking bag. The three main causes of bag failure are:[6]

- Abrasion due to particles that wear the fabric, stresses caused by cleaning, and contact with internal structural supports during the cleaning cycle or normal operation.
- Chemical degradation of the fabric, usually due to incorrect fabric selection or a change in process emissions after exhaust gas characterization has been completed.
- Exceeding the temperature limit of the fabric.

Table 7.4 lists popular fabrics and their operating characteristics.

Even if leakage does not occur, effective filter life can be shortened by factors such as high humidity or inherently sticky particles. These factors can cause dust to adhere to the filter and not be removed during the cleaning cycle. Proper fabric selection or pretreatment of the exhaust gas stream can solve some problems; otherwise, a different type of air cleaner should be specified.

One convenient feature of all filters is that the pressure drop as air passes through them can be easily measured and used as an indicator of filter clogging (Figure 7.2a). The pressure drop increases during operation; when it reaches the design limit, the filters are cleaned or replaced to maintain adequate airflow through the system (Figure 7.2b). In some cases, however, relying only on pressure drop readings to judge filter performance can be misleading. A small hole in the filter may result in a pressure loss that is within acceptable limits even though contaminants are passing through the filter (Figure 7.2c). For this reason, periodic air cleaner efficiency tests or air samples collected at the stack discharge are needed to detect leaks.

Electrostatic Precipitators

Electrostatic precipitators (Figure 7.3) charge particles by means of an electric field that is strong enough to produce ions that adhere to the particles. The charged particles are then collected in a separate section with a weaker electric field that causes the particles to migrate toward and stick to the electrode with the opposite charge.

Precipitators find greatest use in systems where gas volume is large and high collection efficiency for small particles is needed. Precipitators have a relatively low pressure drop and can withstand high temperatures. One disadvantage of electrostatic devices, however, is that they can gradually lose collection efficiency between cleaning cycles if not properly maintained. There is no convenient way to monitor collection efficiency in order to determine when cleaning is needed. Precipitators are also not suited when the exhaust gas is flammable or explosive.

In very small systems, such as portable welding fume or oil mist separators, or for general HVAC system air cleaning, precipitators are

Table 7.4 Baghouse Fabric Selection

Fabric Characteristic	Cotton	Wool	Polyamid (Nylon)	Poly-propylene	Polyester	Aramid	PTFE (Teflon)
Recommended Continuous Service Temperature:							
Dry Heat, °F	180	200	200	200	270	400	425
Wet Heat, °F	180	190	200	200	200	350	425
Resistance to:							
Flex/Abrasion	Good	Fair	Good	Good	Good	Good	Fair
Alkalies	Good	Poor	Good	Excellent	Fair	Good	Excellent
Oxidizing Agents	Fair	Fair	Fair	Good	Good	Poor	Excellent
Acids	Poor	Good	Poor	Excellent	Fair	Fair	Excellent
Organic Solvents	Very Good	Very Good	Very Good	Excellent	Good	Very Good	Excellent
Mildew resistant (without treatment)?	No	No	No Effect	Excellent	No Effect	No Effect	No Effect
Supports combustion?	Yes	No	Yes	Yes	Yes	No	No

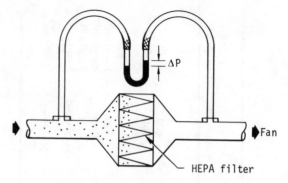

(a) Low pressure drop with clean filter

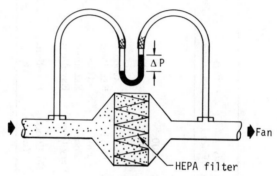

(b) High pressure drop with dirty filter

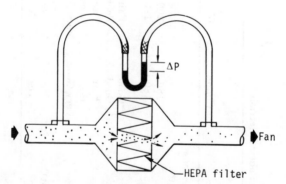

(c) Moderate pressure drop indicates
      leaking filter is working properly

Figure 7.2 Pressure drop tests alone may not
indicate whether a filter is operating properly.
Leak tests to determine removal efficiency are
needed for some highly toxic contaminants.

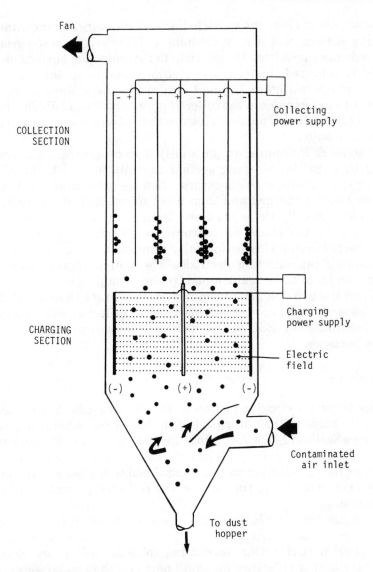

Figure 7.3 Simplified cross-sectional view of an electro-static precipitator.

sometimes called electronic air cleaners. These devices usually operate at a lower charging voltage (12,000 volts versus 45–70,000 volts for large units).

All dusts are characterized by *resistivity,* the resistance of a particle to accepting and holding an electric charge. Extremely high or low values of resistivity make difficult the use of an electrostatic precipitator (ESP).

If resistivity is too low, the particle loses all its charge as it contacts the collecting surface, and it may not adhere. If resistivity is too great, the charges do not move from the particle to the collection surface after the particle is collected. Instead, the charge remains on the particles (a positive charge in Figure 7.3) and diminishes the voltage gradient between the two electrodes in the collecting portion of the ESP. This reduces the rate of migration of particles toward the collecting surface and causes efficiency to suffer.

In some ESP designs, an unusually high concentration of particles moving through the charging section can interfere with the effective charging of all particles. This occurs when particles near the discharge electrode become charged and then repel subsequent ions from the discharge electrode. Particles that are farther away do not receive any charge and are, therefore, not captured in the collecting section.

Collected material is periodically removed by rapping, or mechanically shaking, the collecting electrodes. The material falls away in pieces that are too large to be reentrained by the exhaust gas.

When the exhaust gas contains particles larger than about 20 μm, it may be economical to install a mechanical collector such as a cyclone before installing the ESP so the ESP can be sized to remove only the smaller particles.

Cyclones

Cyclones impart a circular motion to the exhaust gas that causes particulates to move to the outer part of the airstream where they impact the cyclone walls (Figure 7.4). Since air velocity is low at the wall, the particulates drop down the wall into the collection hopper at the bottom. As long as the circular air motion is confined to the body of the cyclone and does not extend into the hopper, the collected particles will not be reentrained.

Cyclones may also be operated as wet collectors if a water spray is installed to wet the particles at the inlet. This increases the effective size of small particles, thus increasing collection efficiency. Small cyclones have higher efficiency for small particles than do larger cyclones. This is because the tangential force on the particle increases as the radius of the cyclone decreases.[7] A bank of smaller cyclones operating in parallel (Figure 7.5) is therefore more efficient than one or two large units. This arrangement maintains good collection efficiency in large airflow volume systems. Cyclones are sensitive to varying airflows since a reduction in airflow also reduces inlet velocity and, hence, the tangential force on the airborne particles.

Cyclones have greatest efficiency for particles 5–10 μm and larger. Since efficiency drops off rapidly below this size range, cyclones are often used as the first stage in a multistage air cleaning system.

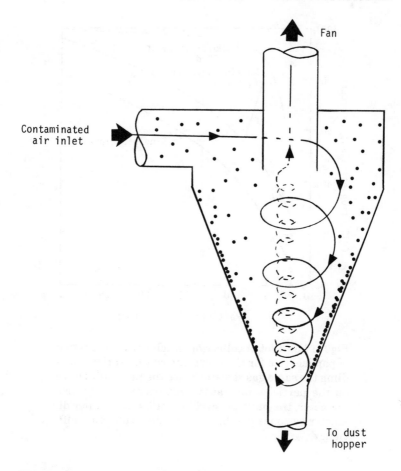

Figure 7.4 Cyclone air cleaner.

Wet Scrubbers

Scrubbers contact particles with water or another liquid and then collect the droplets. To collect extremely fine particles, it is necessary to generate small droplets moving at high speed. Scrubbers can remove particles as small as 0.2 μm; however, the energy (pressure drop) required to generate small droplets and cause adequate contact rises exponentially as the particle size decreases (Table 7.5). Scrubbers that utilize absorption or chemical reaction as a collection mechanism are also widely used for gas and vapor removal.

Scrubbers are a relatively low-cost method of removing small particles in cases where adding liquid to the contaminants does not cause a problem. They are particularly advantageous where the exhaust gas

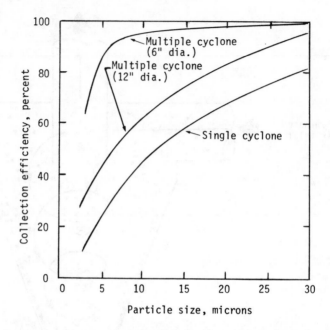

Figure 7.5   Plot of collection efficiency as a function of particle size for three cyclone configurations handling the same gas stream. Since the tangential force on the particles increases with decreasing cyclone diameter, the multiple cyclone unit with smaller diameter cyclones exhibits the highest collection efficiency.

must be cooled; unlike electrostatic precipitators, scrubbers are suitable for flammable or combustible materials.

The use of scrubbers may necessitate treating the waste slurry before disposal. There is also a tendency for fine particles in the recycled scrubbing liquid to erode the spray nozzles, gradually enlarging the opening and increasing droplet size, which reduces collection efficiency.

Table 7.5   Scrubber Pressure Drop vs. Collection Efficiency

Scrubber Type	Typical Pressure Loss (in. H_2O)	Collection Efficiency
Low Contact Power	6	95% of particles >5 μm
Moderate Contact Power	10–14	90% of particles >2 μm
High Contact Power	40–60	90% of particles >0.5 μm

Source: Reference 6.

Chemical imbalances in the scrubbing liquid can also cause formation of precipitates which may plug nozzles.

Scrubbers can collect both particles and gases/vapors but such a combination unit will usually not be highly efficient for extremely small particles. In order to collect fine particles, the spray must be broken into small droplets that travel at high velocity to achieve contact.

The residence time in these units is low since velocity is high. For good gas/vapor removal, a long residence time is required. Since these two collection mechanisms—low residence time and high velocity—work against each other, it is often not economical to build a high energy scrubber for fine particles with an adequate residence time for gas/vapor removal.

Scrubber Designs

Many different scrubber designs are commercially available. The simplest is a spray chamber; the liquid flowrate and pressure determine droplet size. Baffles can be added to the chamber to change airflow direction and capture some particle-containing liquid droplets by impaction (Figure 7.6).

To add more energy to the system, thereby generating smaller droplets and promoting contact, the nozzle pressure can be increased up to 300–450 psig. Combination fan/scrubber units, which use the centrifugal force of the rotating fan blade to promote mixing, are available. Also popular are venturi units, which use the high velocity and turbulence as air passes through a reduced diameter duct section to generate droplets and mix them with the dust.

Another design is the packed bed scrubber (Figure 7.7) in which the liquid flows over a bed of small ceramic rings or specially-shaped pieces that break up the liquid flow. This allows the particles to impinge on the wetted surface of the packing and be flushed into a sump at the bottom of the unit. However, excess pressure drop can result from heavy dust loading that plugs the packed bed. For vertical separators, as illustrated in Figure 7.7, the liquid is often introduced at the top and flows counter to the gas. However, co-current units are also used and can handle higher gas flows with the same size unit and comparable pressure drop.[6]

Cross-flow packed bed scrubbers are arranged with horizontal gas flow through a bed having the liquid introduced at the top. This can reduce the pressure drop, since the air is not supporting the weight of the liquid as in the vertical scrubber. Horizontal units can be protected against having particles block the upstream surface of the packing by providing separate spray nozzles on the upstream face of the packing to flush away the particles.

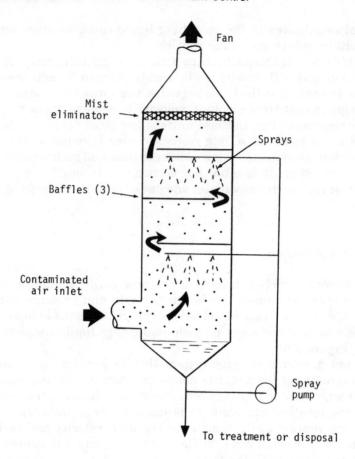

Figure 7.6 Spray chamber with baffles. The baffles assist in collecting particles since wetted particles impact on the baffles.

Packed beds are usually suited for combined particle and gas/vapor removal since the high pressure droplet generation and low residence time characteristics of high energy scrubbers do not apply to packed beds. Packed beds, however, are not highly efficient for very small particles.

Mist Eliminators

All scrubbers may require a mist eliminator, or demister, to remove fine droplets that are entrained in the exhaust gas. Mist eliminators perform three functions:

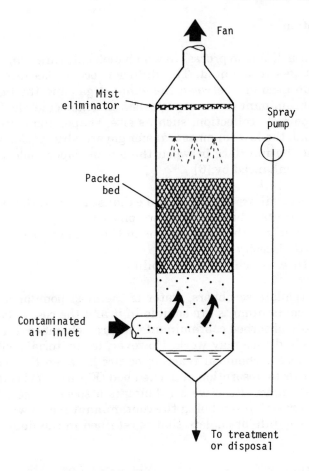

Figure 7.7 Packed bed scrubber for particulates, or gas/vapor contaminants.

- Remove contaminated liquid from the air discharged to the environment.
- Provide additional scrubbing action as the droplets hit the wetted surfaces in the mist eliminator.
- Assist in recovering the scrubbing liquid for reuse.

Mist eliminators are typically mesh pads, louvered plates inclined to the airflow, or a small packed bed.

GAS/VAPOR REMOVAL

Major removal techniques for gases and vapors are absorption, adsorption, and combustion.

Absorption

Absorption is a diffusion process in which molecules are transferred from the exhaust gas to a liquid. The diffusion occurs because there is a concentration gradient between the exhaust gas and the liquid phase. Since the contaminant gas or vapor exists as a molecule, the factors that affect dust particle collection, such as size, shape, and density, do not apply. Instead, the laws of mass transfer govern absorption. Mass transfer occurs at the interface between the gas or vapor molecule and the liquid, and is enhanced by:[6]

- High interfacial area between the exhaust gas and the liquid.
- Turbulent contact between the two phases.
- High solubility of the gas or vapor in the liquid phase.
- Low liquid viscosity.
- Temperature, which affects solubility.

As with particulate scrubbers, water is the most popular absorber because many contaminants are soluble in it and the cost is low.

For easily absorbed contaminants, a spray chamber (Figure 7.6) or another simple device may work. However, for materials with low solubility or where a chemical reaction occurs between the contaminant and liquid prior to absorption, a packed bed (Figure 7.7) is often needed to maximize contact. Reactive scrubbing is a special case of gas-vapor scrubbing. In reactive scrubbing the contaminant reacts with the liquid to form a precipitate or chelate that is retained in the liquid.

Adsorption

Adsorption is the process in which a gas or vapor adheres to the surface of a porous solid material. It occurs when the contaminant condenses into a liquid droplet at an ambient temperature higher than its boiling point. This condensation is due to a catalyzing effect of "active sites" on the surface of the adsorber. The adsorption takes place largely in the internal pore space of the solid (Figure 7.8). For this reason, the effective surface area for adsorption is many times larger than the exterior surface area of the adsorber particles.

Adsorption is reversible since no chemical reaction is involved. This can be an advantage since the contaminant can be recovered if its value warrants the expense of separating it from the adsorper by heating, steam flushing, vacuum treating, or any other method that vaporizes the condensed material. Removing the adsorbate regenerates the adsorbent for further use.

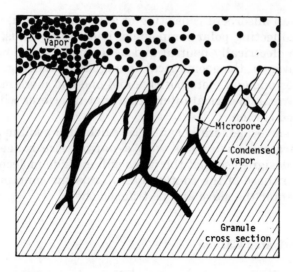

Figure 7.8 Cross-sectional view of activated carbon granule, showing how the vapor is collected in micropores within the granule. (Source: Reference 9)

The capacity of any adsorbent to capture and retain a contaminant depends on:[6]

- Concentration of the contaminant.
- Total available surface area and the number of active sites.
- Size distribution of pores compared to the optimum size for the specific contaminant.
- Tendency of the contaminant to condense at the active sites.
- Competition from other substances in the exhaust gas for available adsorption sites.
- Temperature. Too high a temperature reduces adsorption. Adsorption, like condensation, is an exothermic (heat releasing) process and heat removal may be required to maintain proper temperature.

Adsorbents

There are three major types of adsorbents used in air cleaners:

- Activated carbon, produced from wood, nut shells, or petroleum, is the most common. To remove vapors such as mercury, the activated carbon can be impregnated with iodine or sulfur.

- Silica gel, an adsorbent consisting of amorphous silica, is especially useful in collecting polar compounds. Polar compounds, such as alcohols, are electrically neutral and hence have a zero net electrical charge. However, the geometrical arrangement of charge on the molecule leaves one end more positive and one end more negative. The result is a weak electrical interaction that interferes with collection on activated carbon.
- Molecular sieves, generally synthetic calcium or sodium zeolites (hydrated silicates), have pores with a narrow size range. Molecular sieves are designed so that the pores will exclude molecules that are larger than those of the contaminant. Thus the contaminant molecules can enter the sieve and be trapped there.

Using Activated Carbon

Since activated carbon is the most widely-used adsorbent, it illustrates the application of adsorption devices. If the contaminant has a strong affinity for activated carbon, the carbon bed is essentially 100% efficient until the pores are saturated. Beyond saturation, the contaminant breaks through the carbon bed, first in low concentrations and then at increasingly higher levels as adsorption efficiency drops (Figure 7.9).

Whether a carbon adsorption air cleaner is feasible in a given application depends on the weight of carbon needed to collect the contaminant. The amount of carbon used is a function of the quantity of contaminant in the exhaust stream (concentration × flow rate) plus the relative affinity of the contaminant for adsorption on carbon.[8] In general the less volatile compounds have more affinity for carbon than the more volatile (lower boiling point) materials. High relative humidity (above 50%) and high temperatures decrease the service life of carbon. Depending on the size of the system and carbon usage, the carbon can be discarded after a single use, regenerated at the plant, or returned to the supplier for regeneration.

When you are deciding whether a carbon adsorption air cleaner is feasible, you may need to use bench scale tests to measure carbon service life. A simple equation for estimating carbon service life for organic gases and vapors has been developed.[9] The equation can be used to estimate the time interval from initial installation to the 10% breakthrough point. This is when 10% of the contaminant level entering the carbon bed passes through it. Although 10% breakthrough may be too great for some highly toxic materials, the equation is the best way to estimate carbon service life:

$$t_{10} = \frac{3.85 \times 10^7 \times W_c \times (a + b\, t_{bp})}{C^{2/3} \times M \times Q} \qquad (7.1)$$

Figure 7.9 Contaminants penetrate the carbon adsorption bed as the carbon granules become saturated. (Source: Reference 8)

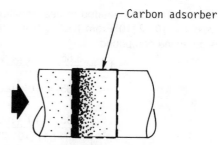

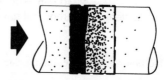

(a) Carbon removal efficiency nearly 100%

(b) Carbon bed nears saturation, breakthrough begins

(c) Saturated carbon no longer removes contaminant

where t_{10} = time until 10% breakthrough, min
W_c = weight of carbon, pounds
a, b = experimental coefficients for different material classes (from Table 7.6)
t_{bp} = contaminant boiling point, °C
C = contaminant concentration entering carbon, ppm
M = contaminant molecular weight
Q = airflow rate, ft³/min

This equation cannot deal with mixtures of contaminants and no provision is made for materials with more than one functional group (a chlorinated alcohol, for example).

Example: Diisopropylamine is used in a chemical manufacturing process. The laboratory hoods where it is heated are ventilated and tests show that the average concentration in the exhaust duct is 37 ppm with an airflow of 400 ft³/min. Estimate the service life of a 100-lb carbon bed before 10% breakthrough occurs.

Answer: The data needed to use Equation (7.1) are: boiling point = 83°C; molecular weight = 101.2.[10] From Table 7.6, a = 0.037 and b = 0.0033, since diisopropylamine is an amine compound.

$$t_{10} = \frac{3.85 \times 10^7 \times W_c \times (a + b \, t_{bp})}{C^{2/3} \times M \times Q}$$

$$= \frac{3.85 \times 10^7 \, (100) \, (0.037 + 0.0033 \, (83))}{37^{2/3} \, (101.2) \, (400)}$$

$$= \frac{3.85 \times 10^7 \, (100) \, (0.274)}{(11.1) \, (101.2) \, (400)}$$

$$= \frac{1.05 \times 10^9}{4.49 \times 10^5} = 2.34 \times 10^3 \text{ min}$$

$$= 2340 \text{ min} = 39 \text{ hr}$$

For a 40-hr work week the carbon will have to be changed about once per week.

Table 7.6 Coefficients for Calculating Carbon Service Life (Equation 7.1)

Solvent Type	Boiling Point Range, (°C)	Coefficients a	b	Examples
Acetates[a]	50 to 190	−0.050	0.0038	Ethyl acetate, butyl acetate
Alcohols[a]	60 to 160	−0.46	0.0071	Methanol, isopropanol, butanol
Alkanes	20 to 200	0.095	0.0022	Pentane, hexane, cyclohexane
Alkyl benzenes	80 to 220	0.12	0.0024	Benzene, toluene
Amines[a]	−10 to 220	0.037	0.0033	Ethylamine, dipropylamine
Ketones	50 to 220	0.034	0.0029	Acetone, diisobutyl ketone
Monochlorides	−30 to 250	0.032	0.0033	Ethyl chloride, 1-chlorobutane
Dichlorides	40 to 250	−0.092	0.0048	Dichloromethane
Trichlorides	60 to 200	−0.080	0.0056	Chloroform, methyl chloroform
Tetrachlorides	70 to 200	0.19	0.0049	Carbon tetrachloride, perchloroethylene

[a]Lower boiling solvents weighed more heavily when determining coefficients.
Source: Reference 9.

Thermal Oxidizers

Thermal oxidation devices use either direct or catalytic oxidation to convert organic compounds into innocuous substances, primarily carbon dioxide and water vapor. The major expense associated with combustion systems is the auxiliary fuel needed to heat incoming exhaust gas and assure complete combustion of contaminants. Since most of the local ventilation systems covered in this book mainly exhaust room air, combustion is often not very cost effective.

Catalytic units are more expensive initially due to the cost of the catalyst, which usually contains platinum, palladium, copper oxide, or manganese oxide. The function of the catalyst is to permit oxidation to occur at a lower temperature than normal, thereby reducing fuel costs. Typical operating temperatures for catalytic units are 600–900°F compared to 1500°F or higher for direct (uncatalyzed) oxidation devices.

Heavy metals, such as lead and mercury, poison catalysts and render them inactive. Other factors that cause loss of activity are overheating due to too high a level of combustibles, and blocking of active sites by particles in the exhaust gas.

Combustion is very useful for processes that release extremely odorous organic vapors and fumes. To reduce auxiliary fuel costs, it may be advantageous to segregate exhaust streams requiring combustion at their source rather than diluting them with less noxious exhaust gas. This will permit use of a smaller oxidation unit.

SUMMARY

Air cleaner selection depends on the physical state of the contaminants, the characteristics of the contaminant and exhaust gas, and the required air cleaner efficiency. Air cleaners often provide the largest single source of resistance in the ventilation system, and may also present the greatest maintenance and operating problems. Careful initial characterization of the contaminants and exhaust gas is vital to a successful system.

REFERENCES

1. Manzone, R.R. and D.W. Oakes, "Profitably Recycling Solvents from Process Systems," *Pollution Engineering* 5, No. 10, 23 (1973).
2. Federal Register 38, No. 66, 8820 (1973).
3. American Conference of Governmental Industrial Hygienists. *Air Sampling Instruments,* 5th Ed. (Cincinnati, Ohio: ACGIH, 1978).
4. American Welding Society. *Fumes and Gases in the Welding Environment* (Miami, Florida: AWS, 1979).
5. Ashe, J.T. "General Recirculation System Sizing in Welding Applications," *Welding J.* 62, No. 1, 81 (1983).
6. Bethea, R.M. *Air Pollution Control Technology* (New York, New York: Van Nostrand Reinhold Company, 1978).
7. Mutchler, J.E. in G.D. and F.E. Clayton, Eds. *Patty's Industrial Hygiene and Toxicology* Vol. I, Chapter 9 (New York, New York: John Wiley and Sons, 1978).

8. Nelson, G.O. and C.A. Harder. "Respirator Efficiency Studies: V. Effects of Solvent Vapor," *Amer. Industrial Hygiene Assoc. J.* 35, No. 7, 391 (1974).
9. Nelson, G.O. and A.N. Correia. "Respirator Cartridge Efficiency Studies: VIII. Summary and Conclusions," *Amer. Industrial Hygiene Assoc. J.* 37, No. 9, 514 (1976).
10. Sax, N.I. *Dangerous Properties of Industrial Materials* (New York, New York: Van Nostrand Reinhold Company, 1975).

CHAPTER 8

Ventilation System Design

After the hoods and air cleaner have been selected, you are ready to design the rest of the ventilation system. Lay out the ducts and decide where to locate the air cleaner and fan. Then size the ducts and fan so that the system works properly and is economical to install and operate.

DESIGN PRIORITIES

Every part of the ventilation system is important, but some components are more important than others. It may be controversial to assign priorities, but it helps to highlight the system components that are more important in determining whether or not a ventilation system will function properly. The priority is:

Hoods

Hoods are the most important part of the system. As discussed in Chapter 6 there are three hood types (Figure 6.7): enclosures that contain contaminants released by a process; capturing hoods that reach out to capture and draw in contaminants generated outside the hood; and receiving hoods that catch contaminants thrown out by the process. The different hood designs have varying airflow requirements. If the wrong hood type is selected or the wrong design airflow into the hood is chosen, the ventilation system may never work as intended, even if the fan and ducts are designed properly.

Air Cleaner

The air cleaner is second in importance during system design. As far as ventilation system design is concerned, the pressure loss contributed by the air cleaner must be overcome by the fan. To choose the proper fan, the resistance caused by the air cleaner must be known. As discussed in Chapter 5 on airflow principles, the air cleaner may be the largest single source of airflow resistance in the entire ventilation system. On a long-term basis, if the originally selected air cleaner is not adequate for pollution control and must be replaced with a more efficient unit exhibiting a higher pressure drop, the fan may then be undersized. This will require installing a new fan or using high energy consumption steps, such as increasing fan rotating speed, covered in Chapter 13.

Fan

Next in importance are the fan and the ducts leading into and out of the fan. The fan generates the suction in the system that draws air in through the hoods. If the fan is inadequate, the airflow through the system will be too low. The ducts before and after the fan may be considered part of the fan, since poor design of these ducts will reduce the fan's ability to move air. Turbulence and uneven flow patterns at the fan inlet reduce the amount of air the fan will move.[3] Luckily, fans have some built-in flexibility; their capacity can be increased by increasing the rotating speed of the fan wheel or blades. Speeding up the fan is the standard remedy for systems with inadequate airflow. Chapter 9 covers types of fans and fan selection principles. Chapter 13 describes how to increase fan output.

Ducts

Ducts have fourth priority in the overall system. Duct diameters and number and type of bends and elbows affect the resistance in the duct network. In multiple-hood systems a design that results in too much resistance in one branch will interfere with airflow through the entire system. The most difficult part of duct design is to choose duct diameters that distribute the air flowing into each hood to meet design airflow criteria in all hoods.

Of course the goal is to design a system with every component working properly and efficiently. Even if they do not fully understand duct and fan design calculations, engineers and safety and health professionals can make a significant contribution to system design by making sure that the initial hood and airflow selections are correct.

FAN AND DUCT DESIGN INFORMATION

The functions of the hoods and air cleaner were emphasized in Chapters 6 and 7. Fans are discussed in Chapter 9, since final fan selection is made only after the system is designed. Information necessary for system design, such as duct diameter and fan size, is summarized below.

Duct Diameter

The magnitude of the pressure losses in the system depends on the square of air velocity through the ducts and other components. Duct velocity, in turn, depends on the diameter of the ducts. Duct diameter is determined according to these guidelines:

- Smaller diameter ducts are less expensive to fabricate and install than larger diameter ducts; however, the resulting higher duct velocities in smaller diameter ducts increase pressure losses in the ventilation system. Thus a larger fan with higher power consumption will be needed for proper system operation.
- One method of designing multiple-hood systems is choosing duct diameters that cause the correct amount of pressure loss to distribute airflow through each hood.
- Systems handling dusts and other solids require a minimum duct velocity to prevent the material from settling in the ducts and plugging them. Table 8.1 lists minimum transport velocities for some common materials.

Although systems handling vapors and gases have no minimum duct velocity criteria, as a rule-of-thumb duct velocities of 2000–3000 ft/min usually result in a good balance between initial duct construction cost and fan operating cost.[4]

Since velocity distribution is more uniform in round ducts than in rectangular ducts, round ducts are used when possible, especially in dust-conveying systems. However, the ACGIH *Manual* [4] contains techniques for calculating equivalent diameters of non-round ducts so that design tables based on round ducts can be used to design the system.

Fan Size

The fan size in a ventilation system must be specified by both volumetric airflow (ft³/min) *and* fan static pressure (inches of water). Together these two parameters specify how much air the fan is to move against the system static pressure, or resistance. Fan static pressure is the amount of static pressure the fan must develop to move the required amount of air through the system. Fan static pressure is defined as the difference between (1) the static pressure loss calculated for the ducts and stack and (2) the velocity pressure in the air entering the fan. The fan static pressure can be calculated as follows:

Table 8.1 Typical Minimum Duct Transport Velocities

Operation	Duct Transport Velocity, ft/min
Barrel filling or dumping	3500–4000
Belt conveyors	3500
Bins and hoppers	3500
Metallizing booth	3500
Melting pot and furnace	2000
Oven hood	2000
Buffing and polishing	
dry dust	3000–3500
sticky dust	3500–4000
Grinding dust	5000
Sandblast dust	4000
Sawdust, wet	4000
Sawdust, dry	3000
Sander dust	2000
Shavings, wet	4000
Shavings, dry	3000
Metal turnings	5000
Lead dust	5000
Welding fumes	1000–3000
Soldering fumes	2000
Paint spray	2000
Grain dust	3000
Cotton dust	3000
Cotton lint	2000

$$\text{FSP} = |\text{SP}_{\text{at fan inlet}}| + |\text{SP}_{\text{stack}}| - \text{VP}_{\text{inlet}} \qquad (8.1)$$

where FSP = fan static pressure, inches of water

$|\text{SP}_{\text{at fan inlet}}|$ = absolute value (disregarding the sign) of static pressure at the fan inlet (representing all losses in the system up to the fan), inches of water

$|\text{SP}_{\text{stack}}|$ = absolute value of static pressure losses in the stack, inches of water

VP_{inlet} = velocity pressure at fan inlet, inches of water

The absolute values of static pressure are used in the calculation because the static pressure on the inlet side of the fan is negative while it is positive on the discharge side of the fan. The sign is not important; only the magnitude of the static pressure is used in Equation 8.1.

Equation 8.1 applies to typical local exhaust systems where only the exhaust stack follows the fan. Therefore, the term SP_{stack} is used in the fan static pressure equation to represent the static pressure (resistance) that the fan must overcome on its discharge side to move the required air volume through the stack. Since the fan may be located anywhere in a local exhaust system, the general version of Equation 8.1 is:

$$FSP = |SP_{inlet}| + |SP_{outlet}| - VP_{inlet} \qquad (8.2)$$

where FSP = fan static pressure, inches of water
       SP = static pressure, inches of water
       VP = velocity pressure, inches of water
       $_{inlet,\ outlet}$ = fan inlet and outlet

Once the fan volume and fan static pressure are known, the type of fan best suited for the system can be selected according to information in Chapter 9.

Design Information

In order to start the design calculations, you will need these items:

- Sketch showing the layout of the machines or operations to be exhausted, the air cleaner and fan location, and connecting ducts.
- Data showing the airflow and hood entry loss factor required for each hood. For many hood designs this information is in the ACGIH *Manual*. For some simple hood types, it is summarized in Table 6.5 (Capture Velocity), Figure 6.23 (Airflow for Capture Hoods), Table 6.6 (Face Velocities for Enclosures) and Figure 5.10 (Hood Entry Loss). Table 8.2 describes hood design features that influence hood entry loss, and lists the figures in other chapters that illustrate these features.
- For dust systems, the minimum duct velocity necessary to avoid deposition of dust in the ducts. This is often included on hood design sheets in the ACGIH *Manual* and is summarized in Table 8.1.
- Pressure drop factors for the air cleaner, if any, from the manufacturer's literature.
- Duct design pressure loss factors for calculating pressure loss in the ducts. Although some of this information is contained in this chapter, more detailed duct design information is in the ACGIH *Manual* and ANSI Standard Z 9.2, "Fundamentals Governing Design and Operation of Local Exhaust Systems."[5]

It also helps to have a ventilation design calculation sheet to keep track of the pressure loss calculations used to determine fan size. Two different calculation sheets are in common use. One equates pressure losses in hoods, duct elbows, and other fittings to the "equivalent length of straight duct" that causes the same pressure loss. The second common

Table 8.2 Hood Entry Loss Values for Typical Hoods

Hood Type	Hood Entry Loss Factor (F_h)	Typical Examples
Low Entry Loss Hood • Open area of hood is large enough to allow smooth airflow pattern into the hood. • Transition from the hood into the duct is tapered to minimize turbulence.	$0.25VP_{duct}$	Figures 5.1, 5.11, 6.2, and 6.24(a)
Moderate Entry Loss Hood • Hood is similar to 'Low Entry Loss Hood' above except that transition into the duct is abrupt (not tapered). • Hood with baffle plates to distribute airflow or other features that interfere with smooth air patterns through the hood. • Hood with restricted open area that results in higher velocity and turbulence in hood.	$0.50VP_{duct}$	Figures 6.11, 6.24(b), 10.5, and 13.7(b)
High Entry Loss Hood • Slot hood where air distribution across the face of the hood is achieved by use of a slot under suction. After entering the slot, the air passes through a plenum chamber into the duct, causing additional entry loss based on duct velocity.	$1.78VP_{slot} + 0.25VP_{duct}$	Figures 6.4, 6.6(left), 6.16, and 6.24(c)

calculation sheet, called the Velocity Pressure Sheet, expresses losses as the number of velocity pressures represented by the loss. Velocity pressure is the amount of kinetic energy in the moving air and is related to velocity by Equation 5.3:

$$VP = \left(\frac{V}{4005}\right)^2$$

where VP = velocity pressure, inches of water

V = velocity, ft/min

Depending on which method of calculating and recording losses is used, design tables based on laboratory tests convert different pressure losses to "equivalent length of duct" or the "number of velocity pressures" lost by that source of turbulence or resistance. Although either calculation method works, this chapter uses the Velocity Pressure Calculation Sheet for the design illustrations since the velocity pressure concept of pressure loss was explained in Chapter 5.

The Velocity Pressure Calculation Sheet (Figure 8.6) is a convenient method of accounting for all losses in a system. Each vertical column represents one branch duct (from hood to the junction with another duct), a section of main duct, or the exhaust stack. Where possible, all pressure losses are expressed in terms of *duct velocity pressure* (Figure 8.6, Section A). If a duct has a slotted hood (a narrow slot with high air velocity is used to distribute airflow across the hood opening), the slot entry loss is proportional to the *slot velocity pressure* (Section B). Some losses are most easily expressed directly in *inches of water* (Section C). Typically this is true for air cleaners, since catalogs state the resistance directly in these units for the design air volume.

Recall from Chapter 5 that the acceleration loss, the energy needed to accelerate room air up to system velocity, must also be accounted for during design. This loss equals "one velocity pressure" corresponding to the highest velocity in the branch duct segment being designed. Where the maximum velocity occurs in the duct, this loss is recorded in Section A (line 12). For a duct with a slotted hood, if the slot velocity exceeds the duct velocity, the acceleration loss is proportional to the slot velocity pressure and is recorded in Section B (line 19). When air in a branch duct enters a main duct, no additional acceleration loss occurs if the air velocity does not increase. This is because the air is already at the duct velocity. If a velocity increase does occur, the incremental additional acceleration loss is added to the main duct design loss.

A blank Velocity Pressure Calculation Sheet is printed at the back of this book to be photocopied for use in following the illustrations or designing other systems. In addition to the tables and figures already discussed, the following duct design information is needed to design the ventilation systems illustrated in this chapter:

- Cross-sectional area of ducts for different duct diameters (Table 8.3). Duct areas are used to calculate duct velocity once the duct diameter and airflow are selected. The duct diameters listed in Table 8.3 are the duct sizes most sheet metal shops can fabricate using standard patterns. Select a diameter from this table rather than using arbitrary duct sizes in designs.

Example: What is the duct velocity in an 8-in.-diameter duct carrying 1200 ft³/min?

Answer: From Table 8.3 the area of an 8-in. round duct is 0.3491 ft².

$$\text{Velocity} = \frac{\text{Volumetric Flow}}{\text{Area}} = \frac{1200 \text{ ft}^3/\text{min}}{0.3491 \text{ ft}^2} = 3437 \text{ ft/min}$$

Table 8.3 Cross-Sectional Area of Ducts

Duct Diameter, in.	Area, ft^2	Duct Diameter, in.	Area, ft^2
1	0.0054	30	4.909
1½	0.0123	31	5.241
2	0.0218	32	5.585
2½	0.0341	33	5.940
3	0.0491	34	6.305
3½	0.0668	35	6.611
4	0.0873	36	7.069
4½	0.1105	37	7.467
5	0.1364	38	7.876
5½	0.1650	39	8.296
6	0.1964	40	8.727
6½	0.2305	41	9.168
7	0.2673	42	9.621
7½	0.3068	43	10.08
8	0.3491	44	10.56
8½	0.3940	45	11.04
9	0.4418	46	11.54
9½	0.4923	47	12.05
10	0.5454	48	12.57
11	0.6600	49	13.10
12	0.7854	50	13.64
13	0.9218	51	14.19
14	1.069	52	14.75
15	1.227	53	15.32
16	1.396	54	15.90
17	1.576	56	17.10
18	1.767	58	18.39
19	1.969	60	19.63
20	2.182	62	20.97
21	2.405	64	22.34
22	2.640	66	23.76
23	2.885	68	25.22
24	3.142	70	26.73
25	3.409	72	28.27
26	3.687	74	29.87
27	3.976	76	31.50
28	4.276	78	33.18
29	4.587	80	34.91

• Velocity pressure as a function of velocity. Velocity pressure is the kinetic energy in the air and is calculated using Equation 5.3. Table 8.4 lists solutions to this equation to facilitate finding duct velocity pressure once you know duct velocity.

Example: What is the velocity pressure corresponding to 3437 ft/min velocity?

Answer: From Table 8.4 the velocity pressure is 0.74 in. of water.

Table 8.4 Velocity Pressures as a Function of Velocity

Velocity, ft/min	Velocity Pressure, in. of water	Velocity, ft/min	Velocity Pressure, in. of water
400	0.010	3500	0.764
500	0.016	3600	0.808
600	0.022	3700	0.853
700	0.031	3800	0.900
800	0.040	3900	0.948
900	0.051	4000	0.998
1000	0.062	4100	1.049
1100	0.075	4200	1.100
1200	0.090	4300	1.152
1300	0.105	4400	1.208
1400	0.122	4500	1.262
1500	0.140	4600	1.319
1600	0.160	4700	1.377
1700	0.180	4800	1.435
1800	0.202	4900	1.496
1900	0.225	5000	1.558
2000	0.249	5100	1.621
2100	0.275	5200	1.685
2200	0.301	5300	1.751
2300	0.329	5400	1.817
2400	0.359	5500	1.886
2500	0.389	5600	1.955
2600	0.421	5700	2.026
2700	0.454	5800	2.098
2800	0.489	5900	2.170
2900	0.524	6000	2.244
3000	0.561	6100	2.320
3100	0.599	6200	2.397
3200	0.638	6300	2.474
3300	0.678	6400	2.554
3400	0.720	6500	2.634

- Duct friction loss data (Figure 8.1) for finding the friction loss in straight ducts expressed in terms of the number of duct velocity pressures lost per 100 ft of duct. To use this figure, you need to know two of these three parameters: duct diameter, duct velocity, or volumetric airflow through the duct.

Example: What is the friction loss in the 8-in. duct carrying 1200 ft³/min if it is 35 ft long?

Answer: On Figure 8.1 locate the intersection of the 8-in. diameter line and the 1200 ft³/min volume line (by interpolation). Then read the friction loss on either vertical axis.

Friction Loss = 2.9 velocity pressures per 100 ft of duct.

As determined earlier, the duct velocity pressure is 0.74 in. of water with a duct velocity of 3437 ft/min. Therefore, the actual friction loss is:

$$2.9 \text{ VP per 100 ft} \times \frac{35 \text{ ft}}{100 \text{ ft}} \times 0.74 \frac{\text{in. of water}}{\text{VP}} = 0.75 \text{ in. of water}$$

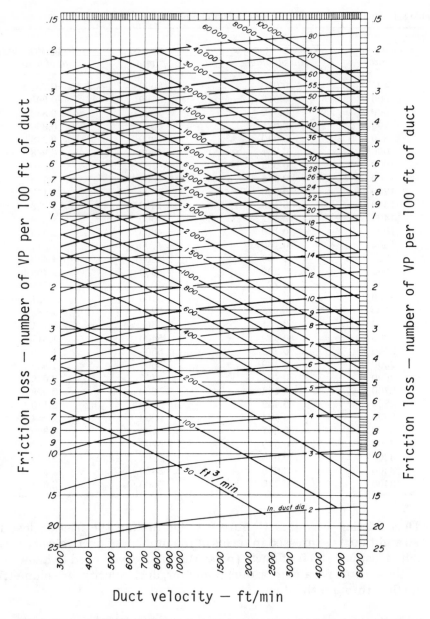

Figure 8.1 Duct friction loss for round ducts expressed as velocity pressure equivalents per 100 ft of duct. (Source: Reference 7 reprinted from the ACGIH *Industrial Ventilation Manual*)

- Pressure loss factors for 90° elbows due to friction and turbulence (Figure 8.2). The magnitude of the factors depends on the ratio of the radius of curvature of the elbow to the duct diameter. A more gradual

R, No. of Diameters	Loss Fraction of VP
2.75 D	0.26
2.50 D	0.22
2.25 D	0.26
2.00 D	0.27
1.75 D	0.32
1.50 D	0.39
1.25 D	0.55

Figure 8.2 Elbow pressure loss for round duct elbows. (Reprinted from the ACGIH *Industrial Ventilation Manual*)

bend (larger radius:diameter ratio) causes less pressure loss than does a tight bend, up to a certain point (2.5D in Figure 8.2). Beyond this point the extra length of duct needed to make the large radius elbow adds more friction loss than is saved by the lower turbulence. For design calculations a 45° bend is considered as 0.5 elbow; a 60° bend is considered 0.67 elbow.

Example: The 8-in. duct described above has two 90° elbows and a 45° bend. If the radius of each elbow is twice the duct diameter, what is the elbow loss?

Answer: From Figure 8.2 the loss per elbow is 0.27 duct velocity pressure. Counting the 45° bend as 0.5 elbow:

$$2.5 \text{ elbows} \times 0.27 \frac{VP}{elbow} = 0.68 \text{ VP}$$

Since the velocity pressure in the duct is 0.74 in. of water, the actual pressure loss is:

$$0.68 \text{ VP} \times 0.74 \frac{in. \text{ of water}}{VP} = 0.50 \text{ in. of water}$$

- Pressure loss factors for branch duct entry into main ducts (Figure 8.3) due to turbulent flow where the two airstreams merge. The entry loss is assumed to occur in the branch duct rather than the main duct and so is recorded in the branch duct column on the Calculation Sheet. The friction loss factor depends on the angle between the main and branch ducts. An entry angle of 30° or 45° is common. Figure 8.3 also shows the preferred branch entry design. The entry occurs at a transition section with the duct diameter after the junction calculated to maintain the same velocity with the total airflow now in the duct. If the duct diameter after the junction is too small and the resultant velocity is significantly higher than in the two ducts before the junction, additional energy will have to be added in the form of static pressure to accelerate the air. Note that branch entry losses are not the same thing as hood entry losses.

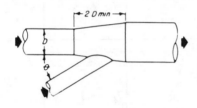

| Angle θ | Loss Fraction of VP |
Degrees	in Branch
10	0.06
15	0.09
20	0.12
25	0.15
30	0.18
35	0.21
40	0.25
45	0.28
50	0.32
60	0.44
90	1.00

Note: Branch entry loss assumed to occur in branch and is so calculated.

Do not include an enlargement regain calculation for branch entry enlargements.

Figure 8.3   Branch duct entry pressure loss. (Reprinted from the ACGIH *Industrial Ventilation Manual*)

Example: The 8-in. duct in Figure 8.3 is a branch duct that enters a main duct at a 30° angle. What is the pressure loss?

Answer: From Figure 8.3 a pressure loss of 0.18 velocity pressure occurs in the 8-in. branch duct. Since the velocity pressure in this duct was previously found to be 0.74 in. of water, the pressure loss is:

$$0.18 \ VP \times 0.74 \ \frac{\text{in. of water}}{VP} = 0.13 \text{ in. of water}$$

These three pressure loss figures are based on "standard air" with a density of 0.075 lb/ft³, which is the density of air at 70°F, 29.92 in. of mercury barometric pressure and 50% relative humidity. If operating conditions cause a significant deviation in air density from this value, the pressure loss through the system will be different than calculated with these design figures. Usually the causes of nonstandard density are temperature and elevation. It is usually easier to calculate the pressure losses through the system assuming standard air and then adjust the fan static pressure value to reflect actual operating conditions as described in Chapter 9. Corrections for density are usually not needed for temperatures between 40° and 100° F and/or elevations of −1000 ft to +1000 ft relative to sea level. For other density corrections, such as high humidity exhaust streams, see the ACGIH *Manual*.

Exhaust Stacks

An exhaust stack on a ventilation system serves two purposes: it helps disperse contaminants in the exhaust stream by discharging the exhausted air above roof level; and it improves fan performance since the uneven velocity distribution at the fan outlet causes a high velocity pressure at the outlet. This higher velocity pressure can result in higher discharge losses if the system has no stack on the fan, as described in Chapter 9.

All systems should have at least a short straight stack on the fan. A high stack discharge velocity (3000 ft/min or higher) helps to disperse contaminants since the air jet action can increase the effective stack height except under severe wind conditions.

If rain entering the stack is a problem, a vertical discharge cap that does not block the stack opening (Figure 8.4a) is effective in keeping

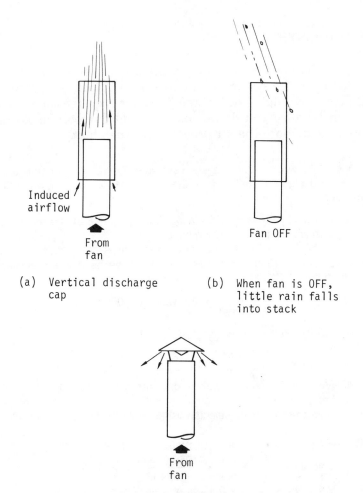

(a) Vertical discharge cap

(b) When fan is OFF, little rain falls into stack

(c) Deflector weathercap is no longer recommended since contaminants are not dispersed

Figure 8.4   Exhaust stack discharge caps.

rain out of the stack (Figure 8.4b) while aiding dispersion with a venturi effect. The old style weather cap deflects exhausted air downward and is no longer recommended (Figure 8.4c). While downward and horizontal fan discharges have been used to prevent rain from entering the stack, such configurations prevent contaminant dispersion and should be avoided. While the fan is operating, rain cannot enter the stack.

Exhaust discharges should be located so that the discharged air will not be drawn back into the building through air inlets for general ventilation.

SINGLE-HOOD SYSTEM

Designing single-hood ventilation systems involves adding up all the pressure losses in the system after choosing the hood type, airflow rate, and duct layout. A single-hood system is easier to design than a multiple-hood system because you do not have to add dampers or adjust duct diameters to distribute the air properly between hoods.

Example: Design the welding bench system shown in Figure 8.5 to operate at standard conditions. Hood design parameters from the ACGIH *Manual* are shown in Figure 6.6 and summarized in Figure 8.5.

Answer: Use the Velocity Pressure Calculation Sheet (Figure 8.6) to record the solution. The entry for each line on the calculation sheet is:

1. Identify the duct before the fan as *Duct* and the exhaust stack as *Stack*.

2. Air Volume = 1050 ft³/min calculated from the hood width and the design information from the ACGIH *Manual* in Figure 8.5.

$$Q = \frac{350 \text{ ft}^3/\text{min}}{\text{ft of width}} \times 3 \text{ ft width} = 1050 \text{ ft}^3/\text{min}$$

3a. Slot Area = 4 in. high $\times \dfrac{\text{ft}}{12 \text{ in.}} \times$ 3 ft wide = 1 ft²

3b. Slot Velocity $= \dfrac{\text{Air Volume}}{\text{Slot Area}} = \dfrac{1050 \text{ ft}^3/\text{min}}{1 \text{ ft}^2} = 1050 \text{ ft/min}$

3c. Slot Velocity Pressure = 0.07 in. of water from Table 8.4.

4. Duct Length = 32 ft; Stack Length = 8 ft.

5. Duct Diameter = 8.5 in., selected using Table 8.3 to give a duct velocity near 2000 ft/min.

6. Duct Area = 0.394 ft² from Table 8.3.

7. Duct Velocity $= \dfrac{\text{Air Volume}}{\text{Duct Area}} = \dfrac{1050 \text{ ft}^3/\text{min}}{0.394 \text{ ft}^2} = 2665 \text{ ft/min}.$

8. Duct Velocity Pressure = 0.45 in. of water (from Table 8.4).

9. Straight Duct Friction Factor = 2.8 VP$_{\text{duct}}$ per 100 ft of duct from Figure 8.1 using 8.5 in. diameter and 2665 ft/min velocity.

10. This system has an elbow with a radius of curvature selected during the preliminary design to be twice the duct diameter. The entry on this line is therefore 2.0. The stack has no elbows so this line is blank.

Figure 8.5 Welding bench exhaust system used for design problem.

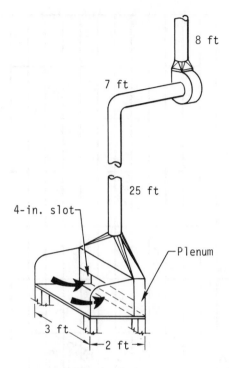

Welding bench design parameters*
Q = 350 ft³/min per foot of bench
  = 350 x 3 = 1050 ft³/min
Slot velocity = 1000 ft/min
Plenum velocity = 1/2 slot velocity
Duct velocity = 1000-3000 ft/min
Entry loss = 1.78 VP_{slot} + 0.25 VP_{duct}
*From Figure 6.6

11. Elbow Pressure Loss Factor = 0.27 VP_{duct} per elbow with R/D = 2.0 from Figure 8.2.

Lines 12–18 (Part A) summarize losses that are calculated in terms of the *Duct Velocity Pressure.*

12. Acceleration Loss = 1.0 VP_{duct}, since the duct velocity exceeds the slot velocity in this system. This is the energy expended in accelerating room air up to duct velocity. Since the stack velocity equals the duct velocity, no additional acceleration loss occurs as the air enters the stack.

13. Hood Entry Loss value is available from the hood design data in Figure 8.5 and equals 0.25 VP_{duct}. This is the turbulent loss as the air moves from the plenum into the duct. Note that the slot loss (1.78 VP_{slot}) is recorded on line 20 since it is based on VP_{slot} rather than VP_{duct}. Since the stack has no hoods, this entry is blank in the stack column.

				Duct	Stack
1	Branch or Main Duct No.				
2	Air volume	ft³/min		1050	1050
3	COMPLETE FOR SLOT HOODS ONLY:				
	a	Slot area	ft²	1.0	
	b	Slot velocity (Line 2 ÷ Line 3a)	ft/min	1050	
	c	VP_{slot} (Table 8-4)	in. H_2O/VP_{slot}	0.07	
4	Duct length	ft		32	8
5	Duct diameter	in.		8.5	8.5
6	Duct area (Table 8-3)	ft²		0.394	0.394
7	Duct velocity (Line 2 ÷ Line 6)	ft/min		2665	2665
8	VP_{duct} (Table 8-4)	in. H_2O/VP_d.		0.45	0.45
9	Straight duct friction factor (Fig 8-1)	VP_d/100ft		2.8	2.8
10	Elbow radius	R/d		2.0	
11	Elbow pressure loss factor (Fig. 8-2)	VP_d/elbow		0.27	

PART A. LOSSES CALCULATED IN UNITS OF VP_{duct}

			Duct	Stack
12	Acceleration loss	1.0 VP_d	1.0	
13	Hood entry loss (Table 8-2, etc.)	VP_d	0.25	0.22
14	Straight duct friction loss (Lines 4 × 9) ÷ 100		0.90	0.22
15	Elbow loss (No. elbows × Line 11)	VP_d	0.27	
16	Branch duct entry loss (Figure 8-3)	VP_d		
17	Other			
18	Subtotal (Part A)	VP_d	2.42	0.22

PART B. LOSSES CALCULATED IN UNITS OF VP_{slot}							
19	Acceleration loss	$1.0\ VP_{slot}$					
20	Slot entry loss (Table 8-2, etc.)	VP_{slot}	1.78				
21	Subtotal (Part B)	VP_{slot}	1.78				
PART C. LOSSES CALCULATED IN UNITS OF "inches of H_2O"							
22	Air cleaner loss (from vendor's data)	in. H_2O					
23	Other	in. H_2O					
24	Subtotal (Part C)	in. H_2O					
SUMMARY OF LOSSES							
25	A. Line 18 × VP_{duct} (from Line 8)	in. H_2O	1.09	0.10			
26	B. Line 21 × VP_{slot} (from Line 3c)	in. H_2O	0.12				
27	C. Line 24	in. H_2O					
28	TOTAL LOSSES (static press.) = 25 + 26 + 27	in. H_2O	1.21	0.10			
29	Governing static pressure	in. H_2O					
30	Corrected air volume	ft^3/min					

$$\text{FSP} = |\text{SP}_{\text{at fan inlet}}| + |\text{SP}_{\text{stack}}| - \text{VP}_{\text{inlet}}$$
$$\text{FSP} = |1.21| + |0.10| - 0.45 = 0.86 \text{ inches of water.}$$

Figure 8.6 Velocity Pressure Calculation Sheet for the welding bench example.

14. Straight Duct Loss Factor is the actual friction loss expressed in terms of duct velocity pressure calculated from lines 4 and 9:

$$\text{Duct:} \frac{32 \text{ ft}}{100 \text{ ft}} \times 2.8 \text{ VP loss per 100 ft} = 0.90 \text{ VP}_{duct}$$

$$\text{Stack:} \frac{8 \text{ ft}}{100 \text{ ft}} \times 2.8 \text{ VP loss per 100 ft} = 0.22 \text{ VP}_{duct}$$

15. Elbow Loss Factor $= 1$ elbow $\times .27 \dfrac{\text{VP loss}}{\text{elbow}} = 0.27 \text{ VP}_{duct}$

16. Branch Duct Entry Loss refers to the loss when a branch duct joins a main duct in a multiple-hood system. This entry is blank for this system.

17. "Other" losses cover miscellaneous losses not included in the calculation sheet. There are none in this system.

18. Subtotal (Part A) summarizes the entries on Lines 12–17 to sum the losses expressed in terms of VP_{duct}.

Lines 19–21 (Part B) summarize losses that are calculated in terms of the *Slot Velocity Pressure*. If there is no slotted hood in the duct, these are blank.

19. Acceleration Loss is blank for the Duct since the duct velocity exceeds the slot velocity for this hood; the acceleration loss is therefore accounted for on Line 12. If the slot velocity were greater than the duct velocity, the 1.0 VP loss would be recorded here and Line 12 would be blank.

20. Slot Entry Loss is available from the hood design data in Figure 8.5 and equals 1.78 VP_{slot}. This is the turbulent loss as the air moves through the narrow slot into the plenum.

21. Summarizes Lines 19 and 20.

Lines 22–24 (Part C) summarize losses that are expressed directly in terms of *inches of water*. Air cleaner loss is a typical example. Inches of water is a convenient unit of measure because it can be measured easily and it is a universally accepted standard, whether in manufacturers' catalogs or architectural specifications.

These lines are blank since this system has no air cleaner or other equipment with losses expressed in "in. of water."

Lines 25–28 summarize the losses in Parts A–C and convert the units to *inches of water* so they can be summed to find total pressure loss.

25. Converts losses expressed in VP_{duct} to inches of water by multiplying Lines 18 and 8:

$$\text{Duct: } 2.42 \text{ VP}_{duct} \times 0.45 \frac{\text{in. of water}}{\text{VP}_{duct}} = 1.09 \text{ in. of water}$$

$$\text{Stack: } 0.22 \text{ VP}_{duct} \times 0.45 \frac{\text{in. of water}}{\text{VP}_{duct}} = 0.10 \text{ in. of water}$$

26. Converts losses expressed in VP_{slot} to in. of water by multiplying Lines 21 and 3c:

$$\text{Duct: } 1.78 \text{ VP}_{slot} \times 0.07 \frac{\text{in. of water}}{\text{VP}_{slot}} = 0.12 \text{ in. of water}$$

27. Zero since Line 24 is blank.

28. Sum of Lines 25–27 to yield the total loss for each duct. This sum represents the suction that is needed in the duct to overcome all pressure losses in drawing the correct air volume.

$$\text{Duct total loss} = 1.21 \text{ in. of water}$$
$$\text{Stack total loss} = 0.10 \text{ in. of water.}$$

29–30. Applies only to multiple hood systems designed for balanced airflow without dampers.

Fan static pressure is the amount of static pressure that the fan must develop to overcome resistance in the system. From Equation 8.1:

$$\text{Fan Static Pressure} = |SP_{\text{at fan inlet}}| + |SP_{\text{stack}}| - VP_{\text{fan inlet}}$$

$$FSP = 1.21 + 0.10 - 0.45 = 0.86 \text{ in. of water}$$

In summary, the system shown in Figure 8.5 should be built using 8.5-in.-diameter ducts and a fan that will deliver 1050 ft³/min of air at 0.86 in. of water fan static pressure. Chapter 9 will explain the types of fans available and how to read a fan rating table in order to select the proper size fan.

MULTIPLE-HOOD SYSTEMS

Designing the ducts for a system with more than one hood is sometimes more complex than designing the ducts for a single-hood system. This is because the design must distribute the total airflow properly between hoods. Before designing a system with more than one hood, review the substances that will be handled at each hood. Some chemicals react when mixed together and can cause corrosion, fire, or explosion. If the system has an air cleaner to remove toxic or nuisance contaminants, check that the air cleaner will not be clogged or saturated by less toxic material from another hood. Sometimes it is advantageous to design several smaller systems rather than a single large system if compatibility is a potential problem.

The general design procedure for multiple-hood systems is the same as for simple systems: choose a diameter for each duct to maintain the design duct velocity and then calculate the pressure loss through the hood, ducts, and other fittings up to the point where another duct joins the duct, or junction. The pressure loss, or static pressure, in each duct that meets at the same junction must be equal at the junction with the design air volume flowing through each duct (Figure 8.7). Look at it another way. The static pressure or suction at the junction in Figure 8.7 is potential energy drawing air through the hoods into the ducts. It is exerted in all directions within the duct and pulls with the same force on both branch ducts. The factor that controls the amount of air that flows through each branch duct is the resistance in that duct. If one duct has a large diameter and the other duct is narrow, then more air will flow through the wider duct. Or if both ducts are the same diameter but one is longer than the other, more air will flow through the shorter duct because the resistance is lower. By creating the correct amount of resistance in each duct, good duct design maintains proper airflow throughout the system. If the resistance is not correct in some ducts, then either

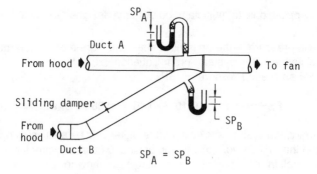

(a)   Balancing with dampers

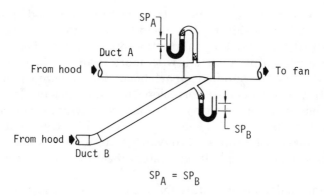

(b)   Static pressure balancing (no dampers)

Figure 8.7   Two methods of designing ducts to obtain proper airflow distribution. When balancing with dampers (a) the diameter of duct B is not important since the sliding damper is adjusted to balance airflow. Static pressure balancing (b) involves selecting the diameter of ducts A and B to give the correct airflow distribution without dampers.

too much or too little air will flow through those ducts and the system may not meet design specifications. The static pressures in each of the ducts meeting at a common junction are equal. This law of physics governs airflow throughout the system. You, as the designer, have three major ways to obtain the proper static pressure loss in ducts:

• Install restrictions, such as adjustable dampers, in branch ducts to create the correct resistance.

- Reduce or increase the diameter of some branch ducts to change the resistance and then recalculate pressure losses. This is repeated until proper balance is achieved.
- Exhaust more air than is required through some hoods to achieve at least the minimum acceptable airflow in all hoods.

The first technique is called "Balancing by Dampers"; the second two methods are used together when "Balancing by Static Pressure." The advantages and disadvantages of the two balancing methods are outlined in Table 8.5. For some ventilation systems, balancing with dampers has an advantage since it provides flexibility in adjusting the system after installation. In addition, less experience is needed to design a workable system. However, dampers are not suitable in systems where dust will collect on the dampers or unauthorized damper adjustments can upset proper balance. The static pressure balancing method (no dampers) results in a system that works well if properly designed. However, the system cannot be easily modified; it is relatively inflexible. For example, if a change in process or configuration is required, moving ducts may unbalance the system.

Balancing with Dampers

Adjustable dampers (also called blast gates) can be installed in branch ducts to permit balancing after the system is installed and operating.

During design, the system layout is reviewed to select the branch of greatest resistance, which is the path from a hood to the fan that has the greatest pressure loss. Once the branch of greatest resistance is chosen, the pressure drop through that path is calculated starting at the hood and working toward the fan. At each point where a branch duct enters the system, the volume of air from the branch duct is added to the volume already in the path of greatest resistance. This new volume is then used for duct sizing and pressure drop calculations. There is no need to calculate the pressure drop through the branch ducts entering the path of greatest resistance. The branch duct diameters are chosen to give the desired velocity, and dampers are used to produce the additional pressure loss needed to balance the system.

Selecting the path of greatest resistance is the key step in the design. If you choose the wrong path and design the system, then the actual path of greatest resistance will never have sufficient airflow, even with all dampers fully open. For typical systems the branch of greatest resistance often leads to the hood most remote from the fan; contains an air cleaner in the branch duct itself; has a high duct velocity; or has a hood with high entry loss caused by slots or small orifices. If there is any doubt which is the branch of greatest resistance, calculate the pressure drop through all combinations until you find the greatest resistance.

Table 8.5   Balancing With Dampers vs. Static Pressure Balancing

Balancing With Dampers		Static Pressure Balancing	
Advantages	Disadvantages	Advantages	Disadvantages
1. Greater flexibility for future changes to system.	1. Unauthorized adjustment of dampers may ruin the protection the system is designed to provide.	1. Unauthorized changes in air volume are difficult to make.	1. Authorized changes in air volumes (such as to correct design errors) are very difficult to make.
2. Correction of improperly selected exhaust volumes is possible after installation.	2. Partially closed dampers may erode or corrode, upsetting balance.	2. Minimum corrosion and erosion problems as compared to corrosion of dampers.	2. If initial estimate of needed airflows was incorrect, duct revision may be needed to correct problem.
3. Design calculations relatively simple.	3. Wrong selection of "branch of greatest resistance" during design might not be discovered until the installed system does not work properly.	3. Wrong choice of "branch of greatest resistance" will be obvious during design.	3. Design procedure is more complex and time consuming.
4. Balance can be achieved with minimum airflow.	4. The job of balancing a complex system after installation may be difficult.		4. Total airflow may exceed minimum necessary in some cases.
5. Minor deviation from design layout is permissible during installation to avoid obstructions in the plant.			5. Installation of ducts must follow layout exactly to maintain properly balanced system.

Example: For the fertilizer screening and bagging operation in Figure 8.8, design the ventilation system with dampers for air balance. Bulk loads of fertilizer pellets are dumped into a hopper from a platform over a vibrating screen that separates the oversize pellets. The oversize pellets drop into a waste barrel while the remaining fertilizer drops into a product hopper for bagging.

Answer: The system consists of 4 separate hoods: dumping platform; screen; bagging; and barrel. Design criteria for each are shown in Figure 8.9. Since transport velocities for different dusts vary, confirm the minimum duct velocity recommendations in Figure 8.9 with Table 8.1.

A schematic of the system, showing dimensions and summarizing the hood design criteria, is shown in Figure 8.10.

By examining the system layout, you will find that the path from the final bagging hood to the fan and stack (Points A–F), which is the longest duct run, is therefore the path of greatest resistance. Although the path from the waste barrel to the stack, path A–G, is almost as long, the air volume from this hood is considerably lower than A–F and so will not cause as much resistance. It will be easier to adjust the pressure loss through duct B–G by increasing the duct size than with duct A–B which carries more air.

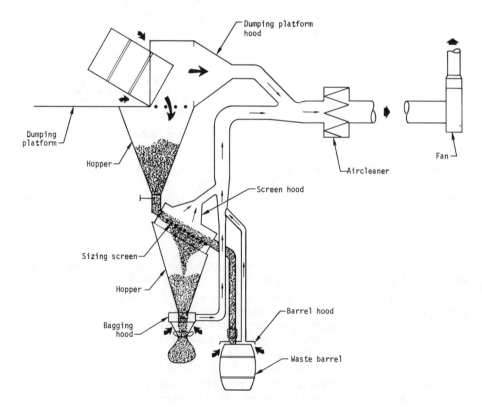

Figure 8.8 Fertilizer sizing and bagging operation used to illustrate the balancing-with-dampers design methods.

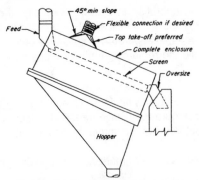

Q = 200 cfm/sq ft through hood openings, but not less than
50 cfm/sq ft screen area. No increase for multiple decks
Duct velocity = 3500 fpm minimum
Entry loss = 0.50 VP

(a) Screen Hood (Design Plate VS-307)

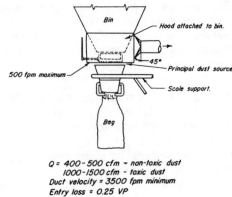

Q = 400 - 500 cfm - non-toxic dust
1000 - 1500 cfm - toxic dust
Duct velocity = 3500 fpm minimum
Entry loss = 0.25 VP

(b) Bag Filling Hood (Design Plate VS-301)

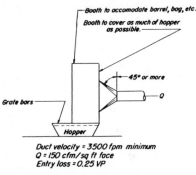

Duct velocity = 3500 fpm minimum
Q = 150 cfm/sq ft face
Entry loss = 0.25 VP

(c) Bin & Hopper Hood (Design Plate VS-304)

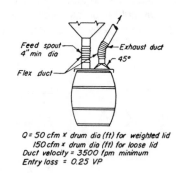

Q = 50 cfm × drum dia (ft) for weighted lid
150 cfm × drum dia (ft) for loose lid
Duct velocity = 3500 fpm minimum
Entry loss = 0.25 VP

(d) Barrel Filling Hood (Design Plate VS-303)

Figure 8.9   Design data for hoods in the fertilizer sizing and bagging system. For clarity, these diagrams are excerpted from the following design plates in the ACGIH *Manual* (Chapter 5): (a) VS-307; (b) VS-301; (c) VS-304; and (d) VS-303.

Use the Velocity Pressure Calculation Sheet (Figure 8.11) to calculate the diameter and airflow through each of the three hoods other than the bagging hood. The idea is to use trial and error to pick the available duct diameter that gives the required duct velocity with as close to the design airflow as possible.

For ducts (G–B) and (H–C) use 3-in. diameter ducts to minimize the possibility of plugging. They have a cross-sectional area of 0.0491 ft² (Table 8.3).

$$\text{Airflow} = \text{Velocity} \times \text{Area}$$
$$= 3500 \text{ ft/min} \times 0.05 \text{ ft}^2 = 175 \text{ ft}^3/\text{min}$$

This is slightly more than the 150 ft³/min required by the design criteria but by using 3-in. ducts rather than smaller sizes the possibility of plugging is reduced.

For duct (I–D) use an 8-in. duct with an area of 0.349 ft².

$$\text{Velocity} = \frac{1350 \text{ ft}^3/\text{min}}{0.349 \text{ ft}^2} = 3868 \text{ ft/min}$$

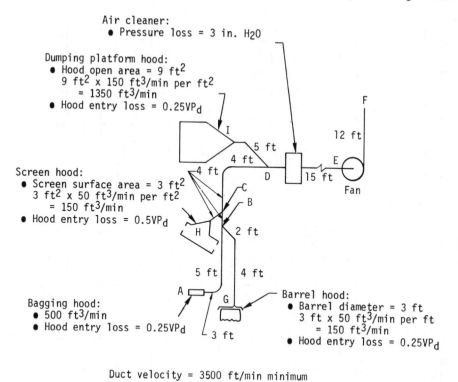

Air cleaner:
- Pressure loss = 3 in. H₂0

Dumping platform hood:
- Hood open area = 9 ft²
  9 ft² x 150 ft³/min per ft²
  = 1350 ft³/min
- Hood entry loss = 0.25VP_d

Screen hood:
- Screen surface area = 3 ft²
  3 ft² x 50 ft³/min per ft²
  = 150 ft³/min
- Hood entry loss = 0.5VP_d

Bagging hood:
- 500 ft³/min
- Hood entry loss = 0.25VP_d

Barrel hood:
- Barrel diameter = 3 ft
  3 ft x 50 ft³/min per ft
  = 150 ft³/min
- Hood entry loss = 0.25VP_d

Duct velocity = 3500 ft/min minimum
Elbow radius/duct diameter = 2.0

Figure 8.10 Schematic layout of the ventilation system to exhaust the operation illustrated in Figure 8.8.

Now select the diameter and calculate the pressure loss through each segment of the path of greatest resistance main duct (A–F). This is done in the same manner as the single duct welding hood system illustrated in Figure 8.6 with one exception: for each duct segment, the airflow in the branch duct entering the main duct is added to the flow already in the main duct. In order to maintain the desired duct velocity with the larger volumetric airflow, a new main duct diameter is calculated for each segment. Each entry in Figure 8.11 is explained below for main duct segments (A–B) and (B–C); the entries for the other main duct segments are calculated in the same way:

1. Identify each segment.
2. Air Volume is 500 ft³/min in (A–B) from the design schematic (Figure 8.10). Air volume is 500 + 175 = 675 ft³/min in duct (B–C) after the airflow from (G–B) enters the main duct.
3. Does not apply since there are no slot hoods involved.
4. Duct Length from Figure 8.10.
5. Duct Diameters are chosen to give the duct velocity of 3500 ft/min.

$$\text{Area} = \frac{\text{Air Volume}}{\text{Air Velocity}}$$

$$\text{Area}_{(A-B)} = \frac{500 \text{ ft}^3/\text{min}}{3500 \text{ ft/min}} = 0.143 \text{ ft}^2$$

#	Branch or Main Duct No...		G-B	H-C	I-D	A-B	B-C	C-D	D-E	E-F
1										
2	Air volume	ft³/min	175	175	1350	500	675	850	2200	2200
3	COMPLETE FOR SLOT HOODS ONLY:									
a	Slot area	ft²								
b	Slot velocity (Line 2 ÷ Line 3a)	ft/min								
c	VP_{slot} (Table 8-4)	in. H_2O/VP_{slot}								
4	Duct length	ft				8	4	8	15	12
5	Duct diameter	in.	3.0	3.0	8.0	5.0	5.5	6.5	10.0	10.0
6	Duct area (Table 8-3)	ft²	0.0491	0.0491	0.349	0.1364	0.1650	0.2305	0.5454	0.5454
7	Duct velocity (Line 2 ÷ Line 6)	ft/min	3564	3564	3868	3667	4091	3687	4034	4034
8	VP_{duct} (Table 8-4)	in. H_2O/VP_d				0.84	1.05	0.84	1.00	1.00
9	Straight duct friction factor (Fig 8-1)	VP_d/100ft				5.4	4.7	3.8	2.3	2.3
10	Elbow radius	R/d				2		2		
11	Elbow pressure loss factor (Fig. 8-2)	VP_d/elbow				0.27		0.27		
	PART A. LOSSES CALCULATED IN UNITS OF VP_{duct}									
12	Acceleration loss	1.0 VP_d				1.0				
13	Hood entry loss (Table 8-2, etc.)	VP_d				0.25				
14	Straight duct friction loss (Lines 4 × 9) ÷ 100	VP_d				0.43	0.19	0.30	0.35	0.28
15	Elbow loss (No. elbows × Line 11)	VP_d				0.27		0.27		
16	Branch duct entry loss (Figure 8-3)	VP_d								
17	Other									
18	Subtotal (Part A)	VP_d				1.95	0.19	0.57	0.35	0.28

PART. B. LOSSES CALCULATED IN UNITS OF VP_slot							
19	Acceleration loss	1.0 VP_slot					
20	Slot entry loss (Table 8-2, etc.)	VP_slot					
21	Subtotal (Part B)	VP_slot					
PART C. LOSSES CALCULATED IN UNITS OF "inches of H₂O"							
22	Air cleaner loss (from vendor's data)	in. H₂O				3.0	
23	Other	in. H₂O					
24	Subtotal (Part C)	in. H₂O				3.0	
SUMMARY OF LOSSES							
25	A. Line 18 × VP_duct (from Line 8)	in. H₂O	1.64	0.20	0.48	0.35	0.28
26	B. Line 21 × VP_slot (from Line 3c)	in. H₂O					
27	C. Line 24	in. H₂O				3.0	
28	TOTAL LOSSES (static press.) = 25 + 26 + 27	in. H₂O	1.64	0.20	0.48	3.35	0.28
29	Governing static pressure	in. H₂O					
30	Corrected air volume	ft³/min					

$$FSP = |\ SP_{\text{at fan inlet}}\ | + |\ SP_{\text{stack}}\ | - VP_{\text{inlet}}$$
$$FSP = |\ 1.64 + 0.20 + 0.48 + 3.35\ | + |\ 0.28\ | - 1.0;\ FSP = 5.67 + 0.28 - 1.0 = 4.95 \text{ inches of water.}$$

Figure 8.11 Velocity Pressure Calculation Sheet for the fertilizer sizing and bagging system. Dampers will be installed in the branch ducts for airflow balancing.

so use a 5-in. duct with an area of 0.1364 ft² from Table 8.3.

$$\text{Area}_{(B-C)} = \frac{675 \text{ ft}^3/\text{min}}{3500 \text{ ft/min}} = 0.19 \text{ ft}^2$$

so use a 5.5-in. duct with an area of 0.165 ft² from Table 8.3.

6. Duct Areas were determined above from Table 8.3.

7. Actual Duct Velocity is the airflow divided by area:

$$\text{Velocity}_{(A-B)} = \frac{500 \text{ ft}^3/\text{min}}{.1364 \text{ ft}^2} = 3667 \text{ ft/min}$$

$$\text{Velocity}_{(B-C)} = \frac{675 \text{ ft}^3/\text{min}}{.1650 \text{ ft}^2} = 4091 \text{ ft/min}$$

8. Duct Velocity Pressure from Table 8.4 is 0.84 in. of water for (A–B) and 1.05 in. of water for (B–C).

9. Straight Duct Friction Factor from Figure 8.1 using the duct diameters and velocities is 5.4 VP per 100 ft of duct for (A–B) and 4.7 VP per 100 ft for (B–C).

10. Elbow radius of curvature for this system was chosen as 2.0.

11. Elbow Pressure Loss Factor is 0.27 VP_d per elbow from Figure 8.2.

12. The Acceleration Loss = 1.0 VP in all ducts with a hood. It is zero for ducts without a hood.

13. Hood Entry Loss = 0.25 VP for the hood in (A–B) from the design information in Figure 8.9. Duct (B–C) has no hood.

14. Straight Duct Friction Loss is

$$\text{for (A–B): } \frac{8 \text{ ft}}{100 \text{ ft}} \times 5.4 \text{ VP per 100 ft} = 0.43 \text{ VP}$$

$$\text{for (B–C): } \frac{4 \text{ ft}}{100 \text{ ft}} \times 4.7 \text{ VP for 100 ft} = 0.19 \text{ VP}$$

15. There is one elbow in (A–B) with a radius of curvature equal to two times the duct diameter. From Figure 8.2 each elbow causes 0.27 VP loss.

$$\text{Elbow Loss Factor}_{(A-B)} = 1 \text{ elbow} \times 0.27 \frac{\text{VP}}{\text{elbow}}$$

$$= 0.27 \text{ VP}$$

16. Branch Duct Entry Loss is zero for both ducts since neither is a branch duct entering a main duct.

17. "Other" is blank since there are no special losses to be accounted for.

18. Subtotal (Part A) is the sum of entries in Lines 12–17.

19–21. These are blank because there are no slot hoods.

22. Air cleaner loss for duct (D–E) is 3.0 inches of water as shown in Figure 8.10 but is blank for (A–C) and (B–C).

23. "Other" is blank.

24. Sums entries on Lines 22 and 23.

25. Converts losses expressed as VP_{duct} to inches of water by multiplying Line 18 by Line 8:

$$\text{for (A–B): } 1.95 \text{ VP}_d \times 0.84 \frac{\text{in. of water}}{\text{VP}_d} = 1.64 \text{ in. of water}$$

$$\text{for (B–C): } 0.19 \text{ VP}_d \times 1.05 \frac{\text{in. of water}}{\text{VP}_d} = 0.20 \text{ in. of water}$$

26. Blank since there are no slot hoods.

27. Carries forward from Line 24 the air cleaner loss expressed in inches of water.

28. Total losses: (Sum of Lines 25–27):

$$\text{for (A–B)} = 1.64 \text{ in. of water}$$
$$\text{for (B–C)} = 0.20 \text{ in. of water}$$

29–30. Applies only to multiple hood systems designed for balanced airflow without dampers.

Finally calculate the Fan Static Pressure using Equation 8.1:

$$\text{Fan static pressure} = |SP_{\text{at fan inlet}}| + |SP_{\text{stack}}| - VP_{\text{at fan inlet}}$$
$$FSP = |1.64 + 0.20 + 0.48 + 3.35| + |0.28| - 1.0$$
$$FSP = 5.67 + 0.28 - 1.0 = 4.95 \text{ in. of water.}$$

Thus the system needs a fan rated at 2200 ft³/min and 4.95 in. of water fan static pressure.

Balancing By Static Pressure (No Dampers)

When a system without dampers is desired, the static pressure balancing design method is used. Proper air distribution through each duct is achieved by selecting duct diameters that generate the needed static pressure at the duct junctions with the design flow.

With this design technique the path of greatest resistance through the system is chosen as with the damper design method. Then the hood and duct in the path of greatest resistance are sized to give the desired duct velocity. However, when a junction is reached where another branch joins this duct, the entering duct is designed starting at the hood and working up to the junction. Then the ducts must be balanced by adjusting air volume or duct diameter until, without dampers, the static pressure in each duct at the junction is equal. The static pressure is the suction needed to pull the desired air volume into the hood and duct.

Increasing Volume to Balance Airflow

If the calculated static pressures in ducts meeting at a junction are within 5% of each other, the difference may be neglected. If the differences in static pressures are within 20% of each other, the easiest way to balance the system is to leave the duct diameters as calculated but increase the air volume flowing through the duct with the lower static pressure. Static pressure or pressure loss depends on the square of the air velocity through the duct. Increasing the air volume increases the velocity. Calculate the new airflow by this equation:[2]

$$Q_{\text{adjusted}} = Q_{\text{design}} \sqrt{\frac{\text{SP duct with larger loss}}{\text{SP duct with smaller loss}}} \qquad (8.3)$$

where Q_{adjusted} = corrected (new) volumetric flow through duct with lower resistance, ft³/min

$Q_{design} =$ original volumetric flow rate chosen for the duct with lower resistance, ft³/min

$SP =$ calculated static pressure, inches of water

After increasing the volume using Equation 8.3, recalculate the pressure loss in that duct to make sure that the static pressures of the ducts at the junction do not differ by more than 5 percent.

Example: Two ducts meet at a junction (Figure 8.7b). One duct is designed to carry 500 ft³/min and has a calculated static pressure at the junction of 1.05 in. of water. The second duct is designed for 400 ft³/min and has a calculated pressure drop of 0.9 in. of water. Balance the airflow at this junction.

Answer: If the airflow is not balanced, the flow through the two ducts will not be correct once the system is installed. More than 400 ft³/min will flow through the second duct since it has lower resistance and less than 500 ft³/min will flow through the other duct.

Find the percentage difference between the two static pressures:

$$\frac{1.05 - 0.9}{1.05} \times 100 = \frac{0.15}{1.05} \times 100 = 14.3\%$$

Since it is greater than 5% but less than 20%, use Equation 8.3:

$$Q_{adjusted} = 400 \text{ ft}^3/\text{min} \sqrt{\frac{1.05 \text{ inches of water}}{0.90 \text{ inches of water}}}$$

$$Q_{adjusted} = 400 \sqrt{1.17} = 432 \text{ ft}^3/\text{min}$$

Since this junction is part of a larger ventilation system, continue the design assuming 432 ft³/min as the airflow in the duct that was originally designed with 400 ft³/min.

Redesigning Ducts to Balance Airflow

If static pressure differences exceed 20%, the best technique is to increase pressure loss by decreasing the duct diameters, slot openings, or the elbow curvature in the branch of lower resistance. You can also reduce pressure loss in the branch with higher static pressure by increasing duct sizes. If you balance airflow by adjusting flows with Equation 8.3, the excess air exhausted would be wasteful.

The balancing procedure is continued for each section of main duct and the entering branch ducts until the fan is reached. The fan static pressure is then calculated according to Equation 8.1.

Example: Use static pressure balancing to design the grinder ventilation system in Figure 8.12.

Answer: Start with the hood design information in Figure 8.12, which is taken from the ACGIH *Manual.* Then, using the first two columns in Figure 8.13, calculate the duct diameter and pressure loss in the two ducts (A–C and B–C) that meet at point C. The calculation for both ducts is similar. Check the pressure loss (static pressure) in each of

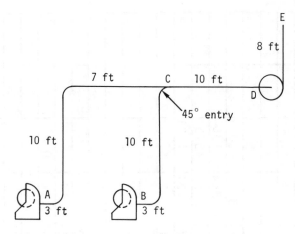

Two 12-in.-diameter grinding wheels

$Q = 300$ ft^3/min

0.65 VP hood entry loss

Minimum duct velocity = 4500 ft/min

Elbow radius/duct diameter = 2.0

Figure 8.12 Pedestal grinder exhaust system to illustrate static pressure balancing.

the ducts to determine whether the system is balanced. Entries according to line number are:

1. Identify the duct segment.
2. Air Volume is 300 ft³/min from Figure 8.12.
3. Does not apply since there are no slot hoods involved.
4. Duct Lengths from Figure 8.12.
5. Duct Diameters are chosen to give the required duct velocity. Choosing 3.5-in. ducts results in duct velocities slightly below the 4500 ft/min minimum but using a 3.0-in. duct would give a velocity over 6000 ft/min.
6. Duct Area from Table 8.3.
7. Duct Velocity $= \dfrac{\text{Volumetric Flow}}{\text{Area}} = \dfrac{300 \text{ ft}^3/\text{min}}{0.0668 \text{ ft}^2}$

$\qquad = 4491$ ft/min

8. Duct Velocity Pressure from Table 8.4 is 1.26 inches of water for a duct velocity of 4491 ft/min.
9. Straight Duct Friction Factor from Figure 8.1, using 3.5-in. duct diameter and 4500 ft/min, is 8.2 VP per 100 ft of duct.
10. Elbow Radius of Curvature for this system was chosen as 2.0.
11. Elbow Pressure Loss Factor is 0.27 VP$_d$ per elbow from Figure 8.2.
12. The Acceleration Loss is 1.0 VP$_d$ for each duct with a hood.
13. Hood Entry Loss for both hoods is 0.65 VP from Figure 8.12.

Initial   Corrected

#	Branch or Main Duct No.		A-C	B-C	B-C	C-D	D-E
1	Branch or Main Duct No.						
2	Air volume	ft^3/min	300	300	324	624	624
3	COMPLETE FOR SLOT HOODS ONLY:						
	a Slot area	ft^2					
	b Slot velocity (Line 2 ÷ Line 3a)	ft/min					
	c VP_{slot} (Table 8-4)	in. H_2O/VP_{slot}					
4	Duct length	ft	20	13	13	10	8
5	Duct diameter	in.	3.5	3.5	3.5	5.0	5.0
6	Duct area (Table 8-3)	ft^2	0.0668	0.0668	0.0668	0.1364	0.1364
7	Duct velocity (Line 2 ÷ Line 6)	ft/min	4491	4491	4850	4575	4575
8	VP_{duct} (Table 8-4)	in. H_2O/VP_d	1.26	1.26	1.47	1.31	1.31
9	Straight duct friction factor (Fig 8-1)	$VP_d/100ft$	8.2	8.2	8.1	5.2	5.2
10	Elbow radius	R/d	2	2	2		
11	Elbow pressure loss factor (Fig. 8-2)	$VP_d/elbow$	0.27	0.27	0.27		
	PART A. LOSSES CALCULATED IN UNITS OF VP_{duct}						
12	Acceleration loss	$1.0\ VP_d$	1.0	1.0	1.0		
13	Hood entry loss (Table 8-2, etc.)	VP_d	0.65	0.65	0.65		
14	Straight duct friction loss (Lines 4 × 9) ÷ 100		1.64	1.07	1.05	0.52	0.42
15	Elbow loss (No. elbows × Line 11)	VP_d	0.54	0.27	0.27		
16	Branch duct entry loss (Figure 8-3)	VP_d		0.28	0.28		
17	Other						
18	Subtotal (Part A)	VP_d	3.83	3.27	3.25	0.52	0.42

PART B. LOSSES CALCULATED IN UNITS OF VP_slot					
19	Acceleration loss	$1.0\ VP_{slot}$			
20	Slot entry loss (Table 8-2, etc.)	VP_{slot}			
21	Subtotal (Part B)	VP_{slot}			
PART C. LOSSES CALCULATED IN UNITS OF "inches of H_2O"					
22	Air cleaner loss (from vendor's data)	in. H_2O			
23	Other	in. H_2O			
24	Subtotal (Part C)	in. H_2O			
SUMMARY OF LOSSES					
25	A. Line 18 × VP_{duct} (from Line 8)	in. H_2O	4.83	4.12	4.78	0.68	0.55
26	B. Line 21 × VP_{slot} (from Line 3c)	in. H_2O				
27	C. Line 24	in. H_2O				
28	TOTAL LOSSES (static press.) = 25 + 26 + 27	in. H_2O	4.83	4.12	4.78	0.68	0.55
29	Governing static pressure	in. H_2O	4.83		4.78	0.68	0.55
30	Corrected air volume	ft^3/min		324		

$$Q_{corr.} = Q_{initial} \sqrt{\frac{SP_{governing}}{SP_{initial}}} = 300 \sqrt{\frac{4.83}{4.12}} = 324\ \text{ft}^3/\text{min.}$$

$FSP = |SP_{inlet}| + |SP_{stack}| - VP_{inlet}$
$FSP = |4.83 + 0.68| + |0.55| - 1.31 = 4.75$ inches of water.

Figure 8.13 Velocity Pressure Calculation Sheet for the pedestal grinder system.

14. Straight Duct Friction Loss:

$$\text{for (A–C): } \frac{20 \text{ ft}}{100 \text{ ft}} \times 8.2 \text{ VP per 100 ft} = 1.64 \text{ VP}$$

$$\text{for (B–C): } \frac{13 \text{ ft}}{100 \text{ ft}} \times 8.2 \text{ VP per 100 ft} = 1.07 \text{ VP}$$

15. Elbow Loss is calculated as follows:

$$\text{for (A–C): 2 elbows} \times 0.27 \frac{\text{VP}}{\text{elbow}} = 0.54 \text{ VP}$$

$$\text{for (B–C): 1 elbow} \times 0.27 \frac{\text{VP}}{\text{elbow}} = 0.27 \text{ VP}$$

16. Branch Duct Entry Loss for a 45° entry is 0.28 VP for (B–C) from Figure 8.3.
17. "Other" is blank.
18. Subtotal (Part A) is the sum of entries for Lines 12–17.
19–21. Blank since there are no slot hoods.
22–24. Blank since this system has no air cleaner with loss expressed in units of inches of water.
25. Converts losses expressed as VP_{duct} to inches of water by multiplying Line 18 by Line 8:

$$\text{for (A–C): } 3.83 \text{ VP}_d \times 1.26 \frac{\text{in. of water}}{\text{VP}_d} = 4.83 \text{ in. of water}$$

$$\text{for (B–C): } 3.27 \text{ VP}_d \times 1.26 \frac{\text{in. of water}}{\text{VP}_d} = 4.12 \text{ in. of water}$$

26–27. Blank as described earlier.
28. Total losses sums Lines 25–27 for each column.
29. Since the static pressure in (A–C) is higher at junction C, it is the governing static pressure at the junction. This means that in order to draw the required 300 ft³/min air volume through duct (A–C), the fan must pull 4.83 in. of water suction at point C. Since duct (B–C) is slightly shorter than (A–C), it has less resistance and does not need as much suction to draw the required 300 ft³/min; however, the higher static pressure must be provided by the fan if (A–C) is to work properly.

The next step is to check whether the junction is balanced. As described earlier in this chapter, the junction is considered balanced if the static pressures, or the sum of pressure losses, in each duct at the junction are equal or at least within 5%. Check the static pressures at this junction:

$$\frac{4.83 - 4.12}{4.83} \times 100 = 14.7\%$$

Since the difference is more than 5% but less than 20%, the easiest way to balance the pressure and thereby generate additional resistance is to increase the airflow through (B–C). Using Equation 8.3:

$$Q_{adjusted} = Q_{design} \sqrt{\frac{SP_{(A-C)}}{SP_{(B-C)}}}$$

$$Q_{adjusted} = 300 \text{ ft}^3/\text{min} \sqrt{\frac{4.83}{4.12}} = 324 \text{ ft}^3/\text{min}$$

This $Q_{adjusted}$ is actually the amount of air that will be pulled through (B–C) when the static pressure at point C is 4.83 in. of water.

30. Corrected Air Volume for (B–C) is 324 ft³/min as described on the preceding page and is used to calculate the duct size and losses for the remaining duct and stack as shown in columns 4 and 5 in Figure 8.13.

Column 3 in Figure 8.13 shows the losses in duct (B–C) recalculated based on an air volume of 324 ft³/min. This recalculation is optional and is shown here to illustrate that the pressure losses at point C are now within the required 5%, after the design airflow in (B–C) is increased:

$$\frac{4.83 - 4.78}{4.83} \times 100 = 1.0\%$$

After duct (C–D) and stack (D–E) are sized as shown in Figure 8.13, the Fan Static Pressure is calculated according to Equation 8.1:

$$FSP = |SP_{at\ fan\ inlet}| + |SP_{stack}| - VP_{at\ fan\ inlet}$$
$$FSP = |4.83 + 0.68| + |0.55| - 1.31$$
$$FSP = 4.75\ inches\ of\ water$$

Select a fan rated at 624 ft³/min and 4.75 in. of water fan static pressure.

This design example illustrated the technique of correcting air volumes when the difference in static pressures at a junction exceed 5% but not 20%. If the difference exceeds 20%, some part of the hood or ducts must usually be redesigned. Using the corrected air volume approach will result in excessive airflow through the "corrected" hood.

To show this, the grinder ventilation system in Figure 8.12 will be redesigned except that duct (A–C) is to be 50 ft long rather than 20 ft as shown in Figure 8.12. To emphasize that the system has been changed, the duct designations will take the form (A′–C′) in the Calculation Sheet used for this example (Figure 8.14).

Example: Redesign the system shown in Figure 8.12 except that duct (A–C) is increased in length from 20 ft to 50 ft.

Answer: Use the Velocity Pressure Calculation Sheet (Figure 8.14) columns 1 and 2 to design ducts (A′–C′) and (B′–C′) as described in the previous example. However, now the total losses in (A′–C′) are increased because the length is greater. Long, small-diameter ducts cause large pressure losses. Due to the need for a high duct velocity to prevent plugging by deposited dust and chips, the diameter of this duct cannot be increased to reduce losses. Duct (B′–C′) has much lower losses because it is shorter.
Check the "balance" at junction C:

$$\frac{7.93 - 4.12}{7.93} \times 100 = \frac{3.81}{7.93} \times 100 = 48\%$$

Because the difference exceeds 20%, duct (B′–C′) will have to be redesigned to increase the losses to a value closer to the 7.93 in. of water loss in (A′–C′). Of course the system

			Initial		Redesigned	
			A'-C'	B'-C'	B'-C'	C'-D'	D'-E'
1	Branch or Main Duct No.						
2	Air volume	ft^3/min	300	300	300	600	600
3	COMPLETE FOR SLOT HOODS ONLY:						
a	Slot area	ft^2					
b	Slot velocity (Line 2 ÷ Line 3a)	ft/min					
c	VP_{slot} (Table 8-4)	in. H_2O/VP_{slot}					
4	Duct length	ft	50	13	13	10	8
5	Duct diameter	in.	3.5	3.5	3.0	5.0	5.0
6	Duct area (Table 8-3)	ft^2	0.0668	0.0668	0.0491	0.1364	0.1364
7	Duct velocity (Line 2 ÷ Line 6)	ft/min	4491	4491	6110	4399	4399
8	VP_{duct} (Table 8-4)	in. H_2O/VP_d	1.26	1.26	2.32	1.21	1.21
9	Straight duct friction factor (Fig 8-1)	$VP_d/100ft$	8.2	8.2	9.8	5.2	5.2
10	Elbow radius	R/d	2	2	2		
11	Elbow pressure loss factor (Fig. 8-2)	$VP_d/elbow$	0.27	0.27	0.27		
	PART A. LOSSES CALCULATED IN UNITS OF VP_{duct}						
12	Acceleration loss	$1.0\ VP_d$	1.0	1.0	1.0		
13	Hood entry loss (Table 8-2, etc.)	VP_d	0.65	0.65	0.65		
14	Straight duct friction loss (Lines 4 × 9) ÷ 100	VP_d	4.10	1.07	1.27	0.52	0.42
15	Elbow loss (No. elbows × Line 11)	VP_d	0.54	0.27	0.27		
16	Branch duct entry loss (Figure 8-3)	VP_d		0.28	0.28		
17	Other						
18	Subtotal (Part A)	VP_d	6.29	3.27	3.47	0.52	0.42

PART B. LOSSES CALCULATED IN UNITS OF VP$_{slot}$								
19	Acceleration loss	1.0 VP$_{slot}$						
20	Slot entry loss (Table 8-2, etc.)	VP$_{slot}$						
21	Subtotal (Part B)	VP$_{slot}$						
PART C. LOSSES CALCULATED IN UNITS OF "inches of H$_2$O"								
22	Air cleaner loss (from vendor's data)	in. H$_2$O						
23	Other	in. H$_2$O						
24	Subtotal (Part C)	in. H$_2$O						
SUMMARY OF LOSSES								
25	A. Line 18 × VP$_{duct}$ (from Line 8)	in. H$_2$O		7.93	4.12	8.05	0.63	0.51
26	B. Line 21 × VP$_{slot}$ (from Line 3c)	in. H$_2$O						
27	C. Line 24	in. H$_2$O						
28	TOTAL LOSSES (static press.) = 25 + 26 + 27	in. H$_2$O		7.93	4.12	8.05	0.63	0.51
29	Governing static pressure	in. H$_2$O		7.93		8.05	0.63	0.51
30	Corrected air volume	ft^3/min						

$$FSP = |SP_{inlet}| + |SP_{stack}| - VP_{inlet}$$
$$FSP = |8.05 + 0.63| + |0.51| - 1.21 = 7.98 \text{ inches of water.}$$

Figure 8.14 Velocity Pressure Calculation Sheet for the revised pedestal grinder system.

would work if a "corrected" air volume was calculated for (B'–C') based on the 7.93 in. static pressure that will be needed to make (A'–C') function. From Equation 8.3:

$$Q_{corrected} = Q_{design} \sqrt{\frac{SP_{duct \ with \ larger \ loss}}{SP_{duct \ with \ smaller \ loss}}}$$

$$= 300 \sqrt{\frac{7.93}{4.12}} = 300 \sqrt{1.92}$$

$$Q_{corrected} = 417 \ ft^3/min$$

To use this approach, the system would draw 117 ft³/min more than is necessary with corresponding increased fan size and power costs.

If dampers were allowed in this system, a damper could be installed in duct (B'–C') and adjusted to provide an artificial pressure loss after startup. However, this system is to be designed without dampers for one or more of the reasons discussed in Table 8.5.

Increasing the loss in (B'–C') is a trial-and-error procedure. Reducing the duct diameter is the most common method to try first. Column 3 in Figure 8.14 shows (B'–C') redesigned using a duct diameter of 3 inches. The major changes affecting pressure loss by line number are:

7. Duct Velocity is increased to 6110 ft/min.

8–9. Both VP_{duct} and Straight Duct Friction Factor are increased since they are proportional to duct velocity squared.

14. Duct Friction Loss is increased since Line 9 is higher.

25 and 28. Total losses are significantly increased, mainly because the VP_{duct} is so much higher.

29. Since the losses in the redesigned (B'–C') exceed the loss in (A'–C'), now (B'–C') is the governing duct. Check the balance at point "C":

$$\frac{8.05 - 7.93}{8.05} \times 100 = \frac{0.12}{8.05} \times 100 = 1.5\%$$

Since it is within 5%, there is no need to make further adjustments. If the difference were greater than 5%, an adjusted air volume could be calculated for the lower loss duct as described in the earlier pedestal grinder system (Figures 8.12 and 8.13).

Now that the ducts at point "C" are balanced, duct (C'–D') and stack (D'–E') can be designed and the Fan Static Pressure calculated as shown in Figure 8.14.

DESIGN PROCEDURE SUMMARY

The four system design examples in this chapter illustrate the different techniques for calculating pressure losses and fan size. The most important point is that a method is needed to identify and record all losses so that hoods will function properly when installed.

It should also be clear from these examples that designing a large system can be time-consuming and tedious, especially when a number of trial-and-error attempts are usually needed to find the best solution. For example, before actually recommending that the system in Figure 8.14 be installed, a prudent designer would check to see whether the fan location and duct layout could be changed to shorten the 50-ft run of duct that causes the high pressure loss. It may even be economical to install two separate systems with short duct runs, particularly if the grinders are not used continuously.

Computers are used to design many large systems. Sophisticated programs capable of balancing systems with up to 100 or more separate ducts have been developed and are used by architect-engineers and others who routinely design complex systems. Limited duct design programs also have been written for some personal computers.[5] Even if a program is not available for a specific computer, the popular spread sheet databases can be set up to calculate the entries in calculation sheets such as the Velocity Pressure Calculation Sheet used in this chapter.

MAKE-UP AIR

Make-up air is air that enters the workroom to replace air exhausted through the ventilation system. A room or plant with insufficient make-up air is said to be "air bound" or "air starved." A ventilation system will not work properly if there is not enough air in the room to exhaust. This means that if the ambient static pressure within the room becomes slightly negative, the fan may not work properly against this additional resistance. Propeller fans used for general ventilation or in large spray booths are especially susceptible to air-bound conditions, since a very slight increase in static pressure decreases their capacity.[4]

More typical signs of insufficient make-up air are:

- Natural stacks and chimneys, which may operate at pressures of 0.01 inches of water draft, that do not work properly. Improperly working stacks and chimneys may cause smoky burner flames and build-up of smoke, carbon monoxide, and other combustion products in the workroom.
- Cold drafts in the workroom caused by cold outside air entering through cracks around doors and windows and other openings. Negative pressures as low as 0.02 inches of water can cause drafts.
- Differential pressure on doors causing doors to slam or be hard to open. Negative pressures of 0.05 to 0.1 inches of water make doors difficult to open.
- Condensation on ceilings and walls in cold areas. In severe cases rain can be drawn in through cracks and will run down inside walls.

Supplying Make-Up Air

Make-up air should be supplied through a planned system (Figure 8.15) rather than through random infiltration. Cold drafts can be eliminated

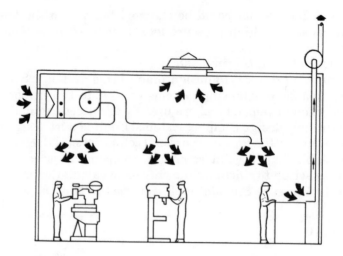

Figure 8.15   Make-up air supply system is important since exhaust air must be replaced.

by heating the air before it is discharged into the workroom. This may seem to be an extra cost but the infiltrated air is also heated as it disperses in the room and lowers the room temperature enough to start the general heating system. Here are some guidelines for make-up air systems:

• The supply rate should exceed the exhaust rate by about 10%. This slight positive pressure in the building helps to keep out drafts and dust. The exception is a situation where no dust or airborne chemicals should travel from the workroom to adjacent offices or other areas. Then a slight negative pressure inside the room is justified.
• The air should flow from cleaner areas of the plant through areas where contaminants may be present and finally to the exhaust system, which removes the contaminated air. Flow should also be from normal temperature areas to high heat process areas. The make-up air supply system can be designed to provide some cooling in the summer in hot process areas.
• Make-up air should be introduced into the "living zone" of the plant, generally 8–10 ft from the floor. This gives the workers the benefit of breathing fresh air and, if the air is tempered (heated or cooled), maximizes the comfort provided by the make-up air.
• The air should be heated in winter to a temperature of about 65°F. The amount of heat needed to temper make-up air is discussed in Chapter 11.

- Make-up air inlets outside the building must be located so that no contaminated air from nearby exhaust stacks or chimneys is drawn into the make-up air system.

Recirculation of Exhausted Air

Recirculation of exhausted air to conserve heat has traditionally been discouraged, especially for systems handling toxic contaminants or high concentrations of any material. Recirculation means that air exhausted from a workplace is cleaned and returned to the work area (Figure 8.16) rather than discharged to the outside and replaced with fresh air that needs cooling or heating.

The problem is that for some very large exhaust systems recirculation seems to be the only practical way of operating the system. Increasing fuel costs and the need to conserve energy also makes recirculation more desirable for many smaller systems.[6] Recirculation may be acceptable in some circumstances provided adequate safety precautions are taken. Precautions include multiple air cleaning systems

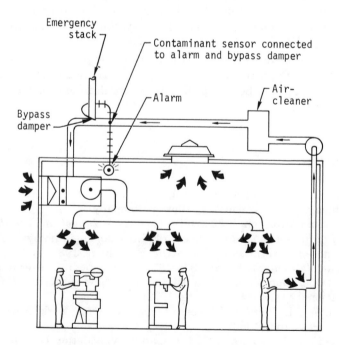

Figure 8.16 If proper precautions are followed, recirculation of exhausted air may be permissible under certain conditions.

installed in series, or automatic sensing devices that warn of air cleaner failure and can divert the recirculated air outdoors if the air cleaner fails.[2]

SUMMARY

This chapter covered the design of local exhaust ventilation systems after the hoods for each pickup point are selected. The design process involves calculating duct diameters to give the desired duct air velocity through the system. The resistance to airflow, or static pressure, through the hoods, ducts, air cleaners, and other components is calculated so that the fan size needed to move the correct amount of air against the system resistance can be determined. Chapter 9 contains guidelines on fan selection.

There are two ways to balance static pressure loss in the separate ducts of a multihood system. One uses adjustable dampers in the ducts to balance the airflow through each hood after installation. The other method relies on proper selection of duct diameters and air volumes to balance the airflow correctly without any dampers. The damper method usually results in a system that is less sensitive to errors in design assumptions and is more flexible to later system changes than a system with no dampers.

An adequate supply of make-up air must be provided to replace the air exhausted through the ventilation system. Although fresh air make-up is preferred, it may be acceptable to recirculate exhausted air under certain conditions.

REFERENCES

1. Trickler, C.J. "Effect of System Design on the Fan," Engineering Letter No. E-4 (Chicago, Illinois: The N.Y. Blower Company).
2. ACGIH Committee on Industrial Ventilation. *Industrial Ventilation—A Manual of Recommended Practice,* 17th Ed. (Lansing, Michigan: American Conference of Governmental Industrial Hygienists, 1982).
3. American National Standard Z 9.2–1971. "Fundamental Governing the Design and Operation of Local Exhaust Systems" (New York, New York: American National Standards Institute, 1972).
4. Hemeon, W.C.L. *Plant and Process Ventilation* (New York, New York: Industrial Press, Inc., 1963).
5. Clapp, D.E., D.J. Groh, and C.M. Nenadic. "Ventilation Design by Microcomputer," *Amer. Industrial Hygiene Assoc. J.* 43, No. 3, 212 (1982).
6. Hama, G.J., "Conservation of Air and Energy," *Michigan's Occupational Health* 19, Spring 1–2 (1974).

7.   Burton, J.R. "Friction Chart," courtesy of Quaker Oats Company, Chicago, Illinois.

CHAPTER 9

Fans

If hoods are the most important component in the ventilation system, the fan and the ducts leading into and out of the fan also rank high in importance. The fan generates the suction in the system that draws contaminated air in through the hoods. If the fan is too small, the airflow will be too low. Fortunately, fan selection does not always have to be perfectly accurate. Fans have some built-in flexibility since their capacity increases with higher fan speeds. Although speeding up the fan is the standard remedy for systems with inadequate airflow, this also increases the fan's power consumption.

The ducts before and after the fan can almost be considered part of the fan itself. These ducts establish smooth airflow into and out of the fan so the fan can do the maximum work moving air. Poor design of these ducts can lead to turbulence and uneven flow patterns at the fan inlet and outlet, and the fan's capacity will be lower than you might expect from the fan size and speed.

This chapter covers fan selection criteria and the different types of fans that are available. The information should be valuable as general background and as a guide in selecting a new fan for a new or existing ventilation system. In many plants the problem is an existing system that does not work properly. Chapter 13, "Solving Ventilation System Problems," reviews steps to improve system performance. Chapter 11 contains tips on lowering the cost of ventilation systems.

FAN AND SYSTEM CURVES

Fan selection involves choosing a fan to match the requirements of the exhaust ventilation system. The fan must move the correct quantity of air against the resistance to airflow caused by friction and turbulence in the system. The relationship between flow rate and resistance for both the exhaust system and the fan can be plotted to help select the proper fan.

Exhaust System Curves

The pressure loss or resistance to airflow through a ventilation system is proportional to the square of air velocity through the hood and ducts. Once a system is designed, and the duct diameters and lengths chosen, the amount of static pressure (suction) that the fan must develop to pull different quantities of air can be estimated and plotted as a *system curve*. Figure 9.1 illustrates a system curve for the welding bench hood system in Figure 8.5. At the design flow rate of 1050 ft³/min the fan must develop 0.86 in. of water fan static pressure. However, to draw 1500 ft³/min, almost 2 in. of water fan static pressure would be needed. Other airflows and fan static pressure values can be read from the curve, which is developed by calculating the required fan static pressures for different airflows as illustrated in Figure 8.6.

Fan Curves

Three curves can be used to describe the performance of a fan. They are the static pressure curve, brake horsepower curve, and mechanical efficiency curve.

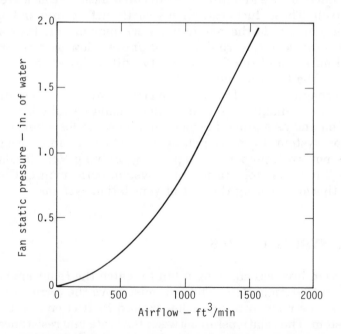

Figure 9.1 System curve for the welding bench exhaust system in Figure 8.5. The static pressure (resistance to airflow) increases with higher airflows through the system.

As discussed later in this chapter, several different types of fans are used in exhaust ventilation systems. Each has different characteristics that make it a good choice for certain applications. The shape of the fan curves is important in selecting the best fan for a ventilation system.

Static Pressure Curve

The quantity of air that a fan will deliver at a given rotating speed depends on the resistance it is working against. As a general rule you expect that the higher the resistance, the less air a fan will move. This is true for all fan speeds that have equal efficiencies. However, the efficiency of the fan can vary as a function of speed. Fan blades are designed for optimum efficiency over various speed ranges. Outside these ranges the fan is especially inefficient and moves less air than expected. All values of fan output and the corresponding static pressure for a given rotating speed can be plotted (Figure 9.2a) as a fan static pressure rating curve.

Fan static pressure is the amount of suction and positive pressure that the fan adds to the ventilation system. It equals the sum of turbulent and friction losses (the total energy needed in the system to move the air) minus the velocity pressure of the air entering the fan inlet (the kinetic energy already in the system). The equation for fan static pressure is:

$$\text{FSP} = |\,\text{SP}_{\text{inlet}}\,| + |\,\text{SP}_{\text{outlet}}\,| - \text{VP}_{\text{inlet}} \qquad (9.1)$$

where FSP = fan static pressure, inches of water
$|\,\text{SP}_{\text{inlet,outlet}}\,|$ = absolute value (disregarding the sign) of static pressure at the fan inlet and outlet, inches of water
VP_{inlet} = velocity pressure of air at the fan inlet, inches of water

For typical industrial exhaust systems where only the stack follows the fan, $|\,\text{SP}_{\text{outlet}}\,|$ in Equation 9.1 actually represents $|\,\text{SP}_{\text{stack}}\,|$. The absolute values of static pressure are used since the static pressure at the inlet is negative while the static pressure at the fan outlet is positive. Since both represent energy needed to overcome resistance to airflow, the signs are not important.

Example: What is the fan static pressure for a system with a static pressure reading of 2.3 in. of water suction on the inlet side of the fan and 0.8 in. of water positive pressure on the discharge side if the average fan inlet duct velocity is 3000 ft/min (Figure 9.3)?

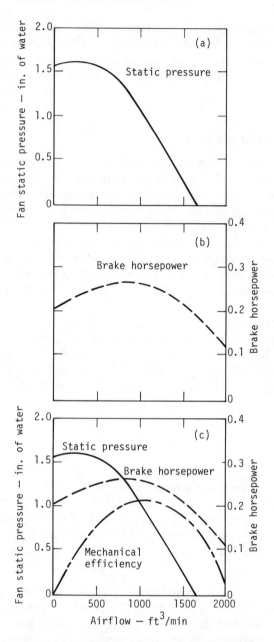

Figure 9.2 Fan curves for a backward inclined blade fan used in exhaust systems: (a) static pressure curve, (b) brake horsepower curve, and (c) the static pressure and brake horsepower curves along with the mechanical efficiency curve.

Answer: The velocity pressure in the duct at 3000 ft/min velocity is 0.56 in. of water (either from Equation 5.3 or Table 8.4).

$$FSP = |SP_{inlet}| + |SP_{outlet}| - VP_{inlet}$$
$$= 2.3 + 0.8 - 0.56$$
$$= 2.5 \text{ in. of water}$$

Brake Horsepower Curve

The amount of electrical power needed to spin the fan depends on the fan's output and the system resistance. It can be plotted as the brake horsepower curve on the fan rating diagram (Figure 9.2b). Brake horsepower is the amount of energy needed to run the fan neglecting the drive losses between the fan and motor. Brake horsepower data are based on manufacturers' tests of their fans following standardized procedures.[1] The actual power consumption will be higher than the brake horsepower rating because of drive losses.

The shape of the brake horsepower curve shows the effect of operating the fan at different points along its static pressure curve. The curve in Figure 9.2 is typical for one type of fan, but other fans have different shaped curves.

Mechanical Efficiency Curve

Mechanical efficiency is a measure of how much energy the fan uses at different points on the static pressure curve. The goal is to choose a fan that is operating near its peak efficiency. The key is to find the correct

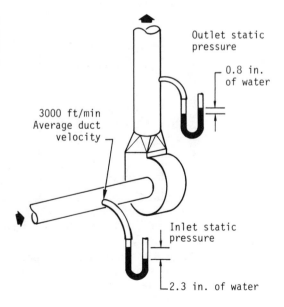

Figure 9.3 Fan static pressure is calculated from the static pressures at the fan inlet and outlet, and the velocity pressure at the fan inlet.

Outlet static pressure

0.8 in. of water

3000 ft/min Average duct velocity

Inlet static pressure

2.3 in. of water

fan size so the peak efficiency coincides with the exhaust system design flow rate and static pressure.

Figure 9.2c shows all three fan curves plotted together as a fan rating curve might appear in the fan manufacturer's literature. Since the mechanical efficiency is a relative measurement there are no units for mechanical efficiency; the shape of the curve indicates its efficiency.

Operating Point

When the fan curves and exhaust system curve are plotted together, the point of intersection of the fan static pressure curve and the system curve indicates the airflow through the system with that fan. The intersection is called the *operating point*. Figure 9.4 shows the system curve for the welding hood system from Figure 9.1 and the fan curves from Figure 9.2. The operating point shows that 1050 ft³/min of air will be drawn through the system with that fan. The brake horsepower can be determined by moving up from the operating point to the brake horsepower curve, then reading the brake horsepower on the right vertical axis. From Figure 9.4, about 0.25 horsepower is the power needed to rotate the fan in that system, neglecting drive losses between fan and motor.

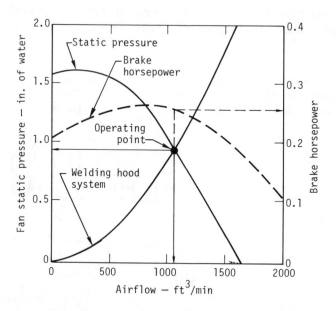

Figure 9.4 Plotting the system curve from Figure 9.1 with the fan curves from Figure 9.2 shows the operating point for the system and fan.

Fan Rating Tables

Every fan has a separate rating curve for each fan rotating speed. Increasing the fan speed moves the curve upward; decreasing the fan speed moves the curve down. Figure 9.5 shows that different operating points for a ventilation system can be achieved by changing the fan rotating speed. Most fan manufacturers make "families" of the same fan design in different sizes with operating curves that cover a range of efficient fan operating conditions. Fan catalogs contain fan rating tables (Figure 9.6) for each size fan constructed from the rating curves. Most fan tables have a shaded portion to indicate the fan to select for maximum mechanical efficiency.

Caution: Fan Rating Data

You can get in trouble using fan rating tables and curves if the air density at the fan is different from the standard density air used to construct the table or if the fan inlet connection does not match the ideal conditions used during the fan tests.

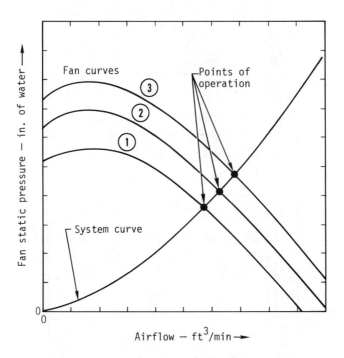

Figure 9.5 Each fan has a separate static pressure curve for each rotating speed.

CFM	1/4" SP RPM	1/4" SP BHP	3/8" SP RPM	3/8" SP BHP	1/2" SP RPM	1/2" SP BHP	3/4" SP RPM	3/4" SP BHP	1" SP RPM	1" SP BHP	1½" SP RPM	1½" SP BHP	2" SP RPM	2" SP BHP	2½" SP RPM	2½" SP BHP	3" SP RPM	3" SP BHP	3½" SP RPM	3½" SP BHP
687	889	0.04	978	0.06	1061	0.08	1218	0.12	1365	0.16	1649	0.28	1896	0.42						
773	957	0.05	1038	0.07	1115	0.09	1261	0.13	1397	0.18	1677	0.31								
859	1028	0.07	1103	0.09	1175	0.11	1310	0.15	1439	0.21										
945	1101	0.08	1172	0.11	1237	0.13	1364	0.17	1486	0.22	1712	0.33	1920	0.45	2114	0.58				
1031	1176	0.11	1241	0.12	1304	0.15	1423	0.21	1537	0.25	1750	0.36	1950	0.49	2139	0.62	2314	0.76		
1117	1253	0.12	1313	0.15	1373	0.17	1486	0.22	1593	0.28	1796	0.41	1988	0.53	2168	0.66	2341	0.81	2504	0.96
1203	1330	0.15	1389	0.17	1445	0.21	1550	0.25	1653	0.31	1845	0.44	2027	0.57	2203	0.71	2368	0.86	2527	1.02
1289	1409	0.17	1464	0.21	1518	0.23	1617	0.29	1714	0.35	1900	0.48	2073	0.62	2240	0.76	2399	0.92	2555	1.08
1375	1488	0.21	1539	0.23	1591	0.26	1688	0.37	1781	0.39	1955	0.52	2122	0.67	2282	0.82	2436	0.98	2584	1.14
1461	1568	0.23	1618	0.27	1666	0.31	1758	0.36	1847	0.43	2015	0.57	2175	0.72	2329	0.88	2476	1.04	2620	1.21
1547	1648	0.27	1696	0.31	1742	0.34	1829	0.41	1914	0.47	2077	0.62	2231	0.78	2378	0.94	2522	1.11	2662	1.29
1633	1730	0.31	1774	0.35	1820	0.38	1904	0.45	1984	0.52	2141	0.68	2289	0.84	2433	1.01	2569	1.18	2704	1.36
1719	1811	0.36	1853	0.39	1896	0.43	1979	0.51	2058	0.58	2207	0.74	2349	0.91	2485	1.08	2621	1.26	2751	1.45
1805	1894	0.41	1934	0.44	1975	0.48	2052	0.56	2128	0.64	2272	0.81	2411	0.98	2545	1.16	2674	1.34	2800	1.54
1891	1974	0.46	2014	0.51	2053	0.54	2130	0.62	2204	0.71	2343	0.88	2477	1.05	2604	1.24	2729	1.43	2850	1.63
1977	2059	0.52	2097	0.56	2134	0.61	2205	0.68	2278	0.77	2414	0.95	2542	1.13	2666	1.32	2789	1.52	2905	1.73
2063	2141	0.59	2176	0.63	2213	0.67	2285	0.76	2352	0.84	2483	1.03	2610	1.22	2728	1.41	2847	1.62	2962	1.83
2149	2222	0.65	2258	0.71	2293	0.74	2362	0.83	2428	0.92	2555	1.11	2677	1.31	2792	1.51	2909	1.72	3021	1.94
2235	2306	0.73	2340	0.77	2375	0.82	2441	0.91	2504	1.01	2627	1.21	2747	1.41	2861	1.62	2971	1.83	3080	2.05
2407	2474	0.91	2505	0.94	2537	0.99	2600	1.09	2660	1.19	2775	1.41	2888	1.62	2995	1.83	3100	2.06	3205	2.31

CFM	4" SP		4½" SP		5" SP		5½" SP		6" SP		6½" SP		7" SP		7½" SP		8" SP		8½" SP	
	RPM	BHP	RPM	BHP	RPM	BHP	RPM	BHP	RPM	BHP	RPM	BHP	RPM	BHP	RPM	BHP	RPM	BHP	RPM	BHP
1289	2699	1.24	2841	1.42																
1375	2730	1.32	2868	1.49	3001	1.68														
1461	2762	1.39	2895	1.57	3024	1.76	3150	1.95	3277	2.16										
1547	2797	1.47	2927	1.66	3052	1.85	3180	2.05	3299	2.26	3417	2.47								
1633	2836	1.55	2963	1.75	3086	1.94	3206	2.15	3323	2.35	3438	2.57	3551	2.79	3660	3.01				
1719	2878	1.64	2999	1.84	3119	2.04	3238	2.25	3356	2.47	3465	2.68	3576	2.91	3684	3.14	3788	3.37	3892	3.61
1805	2922	1.73	3042	1.94	3160	2.15	3273	2.36	3386	2.58	3498	2.81	3603	3.03	3710	3.26	3812	3.51	3916	3.75
1891	2969	1.83	3087	2.04	3201	2.25	3312	2.47	3424	2.71	3530	2.93	3635	3.16	3740	3.41	3843	3.64	3940	3.89
1977	3023	1.94	3137	2.15	3245	2.37	3355	2.59	3462	2.82	3567	3.06	3671	3.31	3771	3.54	3868	3.78	3970	4.04
2063	3075	2.04	3185	2.26	3290	2.48	3398	2.72	3503	2.95	3604	3.19	3707	3.44	3807	3.69	3903	3.94	4000	4.21
2149	3131	2.16	3238	2.39	3340	2.61	3441	2.84	3543	3.08	3644	3.33	3746	3.58	3842	3.83	3938	4.09	4030	4.35
2235	3187	2.28	3290	2.51	3394	2.75	3493	2.98	3590	3.22	3693	3.49	3788	3.74	3881	3.99	3973	4.25	4065	4.51
2407	3306	2.54	3404	2.78	3501	3.02	3599	3.28	3690	3.53	3786	3.81	3879	4.07	3964	4.31	4059	4.61	4143	4.87
2579	3432	2.82	3524	3.07	3617	3.33	3708	3.59	3799	3.86	3888	4.13	3975	4.41	4059	4.68	4147	4.97	4232	5.26
2751	3559	3.13	3651	3.41	3738	3.67	3825	3.94	3910	4.21	3998	4.51	4079	4.78	4164	5.08	4245	5.37		
2923	3692	3.46	3779	3.74	3866	4.03	3947	4.31	4032	4.61	4110	4.89	4191	5.19						
3095	3830	3.83	3912	4.12	3994	4.42	4075	4.72	4154	5.02	4232	5.32								
3267	3969	4.23	4047	4.53	4130	4.85	4206	5.16												
3439	4110	4.66	4191	4.99	4267	5.31														

Selection within tinted area renders most efficient, quietest operation.   BHP shown does not include belt drive losses.

Figure 9.6 Fan rating table for a backward inclined blade fan from a fan catalog. CFM = Flow rate, ft³/min; RPM = Rotating speed, rev/min; SP = Fan static pressure, in. of water; BHP = Brake horsepower. (Source: Reference 2)

Air Density

Fan ratings are developed from tests using "standard air" with a density of 0.075 lb/ft³. This is the density of air at 70°F, 50% relative humidity, and a barometric pressure of 29.92 in. of mercury. When air density varies significantly from this value, corrections to fan ratings are needed. The primary factors affecting density are air temperature and the plant's altitude above sea level. If a wet scrubber (air cleaner) is used, the increased relative humidity may also increase air density. In any case the volume capacity of the fan is not changed, but the static pressure developed by the fan varies with the air density. If the air temperature or the altitude increases, the air becomes less dense. For example, a fan moving 1000 ft³/min of this less dense air is moving less mass than a fan moving the same quantity of "standard air." Consequently, the fan does not have to develop as much static pressure when it is moving less dense air. The horsepower requirement is also lower since less air mass is being moved. Corrections must then be made to the pressure and horsepower ratings using the factors in Table 9.1.

Standard air is usually assumed during system design because the duct friction and other pressure loss design data in Chapter 8 are based on standard air. After the system has been designed and the fan capacity

Table 9.1 Air Density Correction Factors (Altitude and Temperature)

Air Temp., °F	Altitude Above Sea Level									
	0	1000	1500	2000	2500	3000	3500	4000	4500	5000
0	.87	.91	.92	.94	.96	.98	.99	1.01	1.03	1.05
70	1.00	1.04	1.06	1.08	1.10	1.12	1.14	1.16	1.18	1.20
100	1.06	1.10	1.12	1.14	1.16	1.19	1.21	1.23	1.25	1.28
120	1.09	1.14	1.16	1.18	1.20	1.23	1.25	1.28	1.30	1.32
140	1.13	1.18	1.20	1.22	1.25	1.27	1.29	1.32	1.34	1.37
160	1.17	1.22	1.24	1.26	1.29	1.31	1.34	1.36	1.39	1.42
180	1.21	1.26	1.28	1.30	1.33	1.36	1.38	1.41	1.43	1.46
200	1.25	1.29	1.32	1.34	1.37	1.40	1.42	1.45	1.48	1.51
250	1.34	1.39	1.42	1.45	1.47	1.50	1.53	1.56	1.59	1.62
300	1.43	1.49	1.52	1.55	1.58	1.61	1.64	1.67	1.70	1.74
350	1.53	1.59	1.62	1.65	1.68	1.72	1.75	1.78	1.81	1.85
400	1.62	1.69	1.72	1.75	1.79	1.82	1.85	1.89	1.93	1.96
450	1.72	1.79	1.82	1.86	1.89	1.93	1.96	2.00	2.04	2.08
500	1.81	1.88	1.92	1.96	1.99	2.03	2.07	2.11	2.15	2.19
550	1.91	1.98	2.02	2.06	2.10	2.14	2.18	2.22	2.26	2.30
600	2.00	2.08	2.12	2.16	2.20	2.24	2.29	2.33	2.38	2.42
650	2.10	2.18	2.22	2.26	2.31	2.35	2.40	2.44	2.49	2.54
700	2.19	2.27	2.32	2.36	2.41	2.46	2.50	2.55	2.60	2.65
750	2.28	2.37	2.42	2.47	2.51	2.56	2.61	2.66	2.71	2.76
800	2.38	2.48	2.52	2.57	2.62	2.66	2.72	2.76	2.81	2.86

Source: Reference 3.

and static pressure calculated, corrections are made to reflect actual operating conditions. Generally no density corrections are needed for temperatures from 40°F to 100°F or for altitudes from -1000 ft to $+1000$ ft (relative to sea level) because density differences over these ranges are not usually large enough to affect fan performance in industrial exhaust systems.[4]

The most common type of density correction occurs in cases where a system designed to move specific amounts of air at a specific density must now operate with air of different density. In this case you must calculate the correct fan static pressure and horsepower to move the desired air volume.

Example: Although it was designed to operate at sea level, the welding hood system in Figure 8.5 must be relocated to Denver, Colorado (altitude 5000 ft). What size fan will be needed if the original specifications required a fan moving 1050 ft³/min at 0.86 in. of water fan static pressure and a suitable fan had a 0.25 brake horsepower rating at those conditions?

Answer: The fan must move 1050 ft³/min but, since the air density is lower with greater altitude, less air mass will be moved. The result is less resistance and horsepower. From Table 9.1 the correction factor for 5000 ft and 70°F is 1.20.

$$FSP_{actual} = \frac{FSP_{design}}{\text{Density Factor}} = \frac{0.86}{1.2} = 0.72 \text{ in. of water}$$

$$BHP_{actual} = \frac{BHP_{design}}{\text{Density Factor}} = \frac{0.25}{1.2} = 0.21 \text{ brake horsepower}$$

Select a fan from standard fan tables rated at 1050 ft³/min and 0.72 in. of water fan static pressure.

A second type of density correction is applied when you want a system to move a specific volume of air and develop a specific fan static pressure at nonstandard conditions.

Example: Select a fan to move 15,000 ft³/min at 30 in. of water fan static pressure while operating at 200°F and 2500 ft altitude.

Answer: In order to use a fan rating table, the static pressure must be adjusted to the equivalent fan static pressure at standard conditions. From Table 9.1 the correction factor for 200°F and 2500 ft is 1.37.

$$FSP_{equiv} = FSP_{actual} \times \text{Density Factor}$$
$$= 30 \text{ in.} \times 1.37 = 41.1 \text{ in. of water}$$

Select a fan rated at 15,000 ft³/min and 41 in. of water fan static pressure from a standard rating table.

For the dilution ventilation systems discussed in Chapter 1 the air volume itself must be corrected, since the dilution effect is based on air mass rather than air volume. In other words, if calculations show that a rate of 1000 ft³/min is needed for contamination dilution at standard conditions, a higher airflow rate will be needed at operating conditions where the air density is less than 0.075 pounds/ft³. This is explained in Chapter 1.

Poor Fan Inlet Connections

The second thing to remember when using fan rating tables or curves is that the tests used to develop the ratings were conducted under ideal laboratory conditions. Often field conditions do not equal the test conditions; therefore, a fan will not perform as well as the rating table predicts it will.

For the tests, straight ducts are connected to both the fan inlet and outlet (Figure 9.7). This helps assure that the air enters and leaves the fan with minimum turbulence and nonuniform flow. Fan blades are designed to be most efficient when air enters the fan in a straight line. Elbows, fan inlet boxes, or duct junctions near the fan can impart a spin to air entering the fan. If the spin is in the same direction as fan rotation,

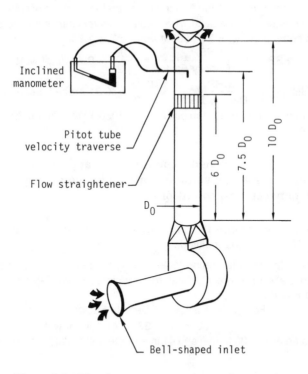

Figure 9.7 The duct arrangement used to test fan performance provides ideal airflow conditions that actual installations may not duplicate. This can reduce fan performance below that expected from the manufacturer's rating table.

the fan blades have to "catch up" to the air before acting on it, so that the amount of air moved will decrease along with energy consumption. If the air spin is opposite to the fan rotation, the output will be reduced, although the power consumption will be higher than expected. Figure 9.8 illustrates how the fan curve can be affected by an elbow or fan inlet box. The same figure also shows that performance can be restored by adding turning vanes inside the box to reestablish straight flow into the fan.[5]

Reduced fan performance due to poor inlet connections is insidious in that it cannot be identified by using the standard pressure measure-

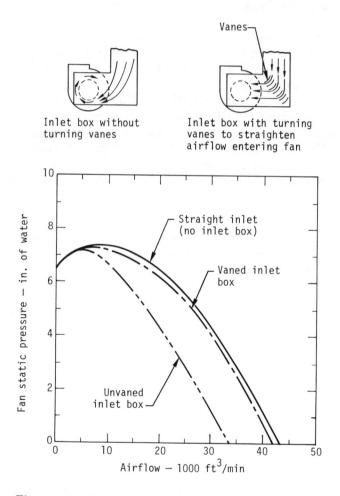

Figure 9.8 The effect on fan performance of a fan inlet box without turning vanes. Addition of turning vanes to straighten airflow restores fan performance. (Source: Reference 5)

ments used to test fan performance.[6] Reduced performance caused by a duct obstruction or a plugged filter can be identified by using the simple pressure tests discussed in Chapter 12. However, poor duct inlet connections do not increase pressure losses in the system; they simply reduce the fan's ability to do useful work on the air. Special testing techniques to measure spinning or uneven airflow into the fan are covered in Chapter 12.

Poor Outlet Connections
Reduce Static Regain

Poor fan outlet connections also have an adverse effect on fan performance. Ideally the fan outlet should be a straight duct length with no elbows or other interference to smooth flow for 5 to 10 duct diameters away from the fan outlet.

The reason for the adverse effect is that the air discharged from a fan outlet does not have uniform velocity distribution (Figure 9.9). Since air has weight, air is thrown out by centrifugal force from the spinning fan wheel. This results in higher velocities at the outer edge of the outlet than at the inner edge. Several duct diameters downstream from the fan outlet, the air velocity returns to near uniform distribution across

Figure 9.9 Uneven air velocity distribution at the fan outlet results in a higher velocity pressure there compared to a location 5–10 duct diameters downstream where air velocity distribution is more uniform.

the duct.[7] The velocity pressure (kinetic energy) in the moving air is proportional to the square of the velocity according to Equation 5.3:

$$VP = \left(\frac{V}{4005}\right)^2$$

where VP = velocity pressure, inches of water

V = air velocity, ft/min

Usually the velocity pressure is calculated from the average duct velocity. For added accuracy, however, the velocity pressure can be calculated using the individual velocity readings across the duct as shown in Figure 9.9:

$$VP = \frac{\left(\dfrac{V_1 + V_2 + \ldots + V_n}{N}\right)^2}{(4005)^2} \tag{9.2}$$

where $V_1, V_2 \ldots V_n$ = individual velocity readings, ft/min

N = number of individual velocity readings

Static Regain

Solving Equation 9.2 at both the fan outlet and a point several duct diameters downstream from the fan outlet shows that the velocity pressure in the system is higher at the fan outlet than downstream from the outlet due to the effect of squaring the high individual velocity readings at the fan outlet. Since no energy was added to the system between these locations, the total energy is constant except for the slight duct friction loss. So the drop in velocity pressure is balanced by a corresponding increase in static pressure as velocity pressure (kinetic energy) is converted into static pressure (potential energy). This phenomenon is known as *static regain* and is important in ventilation work because the magnitude of friction, turbulent, and other pressure losses is directly proportional to the velocity pressure. These losses can be minimized in the exhaust stack by converting as much as possible of the velocity pressure that exists at the fan outlet back to static pressure, before the air reaches elbows or other sources of pressure loss. If you install an elbow immediately after the fan, you will cause turbulence pressure losses in air with high velocity pressure. The result is that losses will be greater than if a short length of straight duct between elbow and fan outlet permitted static regain to occur (Figure 9.10). When elbows cannot be avoided, take advantage of the centrifugal motion of the air at the fan outlet by using

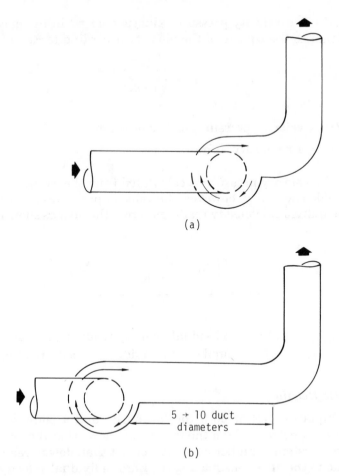

(a)

(b)

5 → 10 duct
diameters

Figure 9.10   An elbow at the fan outlet (a) causes a
higher pressure loss than an elbow located downstream
(b) because the velocity pressure is higher at the elbow
in (a) than at the elbow in (b).

an elbow as in Figure 9.11b rather than the elbow in 9.11a.[8] Often an
elbow can be avoided by rotating the fan housing during installation
(Figure 9.11c).

Exhaust Stacks

Static regain can be carried one step further. In some systems a gradual
taper called an evasé (Figure 9.12) is used to maximize static pressure
regain before the air is discharged from the stack. At the stack discharge,
the air decelerates from the duct velocity to essentially zero velocity in
the ambient environment. Thus energy corresponding to one unit of

Figure 9.11 When elbows cannot be avoided, their direction is important. In (a) the elbow causes the air to change direction from the curving due to centrifugal action. The elbow in (b) takes advantage of this curving airflow while diagram (c) illustrates how rotating the fan housing can eliminate the elbow.

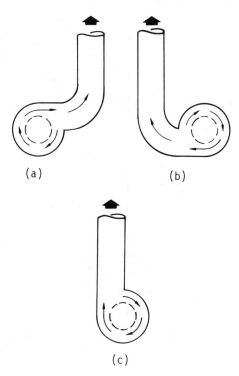

(a)

(b)

(c)

velocity pressure is lost through this deceleration. This energy was added to the system as the air entered the hoods and was accelerated to duct velocity. Slowing the air as much as possible in the stack reduces the magnitude of the deceleration loss. However, a high discharge velocity is advantageous in some ventilation systems since it helps disperse the contaminants in the exhausted air. Of course a system with no stack at all on the fan outlet has a very high deceleration loss. Nonuniform velocity distribution (Figure 9.9) results in a higher velocity pressure at the fan outlet. Figure 9.12 shows that a straight stack does not affect pressure loss in the system; no stack at all on the fan outlet causes a loss of 0.5 VP. Since an evasé permits static regain to occur, the fan size can be reduced in systems with an evasé. Every system should have at least a short straight stack on the fan outlet.

In summary, pressure losses on the discharge side of the fan can be minimized by using static pressure regain techniques to reduce the high velocity pressure at the fan outlet.

CHOOSING THE RIGHT FAN

To choose the proper fan for a ventilation system you need to know the following information:[9]

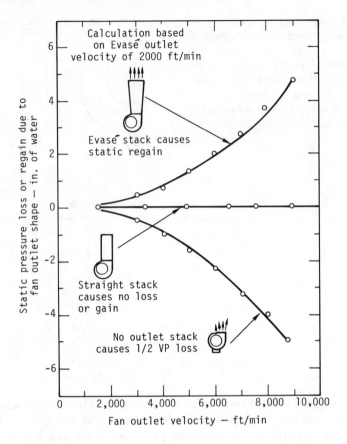

Figure 9.12 The type of fan discharge stack has an effect on static pressure regain and static pressure loss.

- Air volume to be moved.
- Fan static pressure.
- Type and concentration of contaminants in the air. (They affect the fan type and materials of construction.)
- Importance of noise levels as a limiting factor.

Once this information is available, the type of fan best suited for the system can be chosen. There is a variety of different fans available but they all fall into one of two classes: axial flow fans and centrifugal fans.

Centrifugal Fans

Centrifugal fans move air by centrifugal action. Blades on a rotating fan wheel throw air outward from the center inlet at a higher velocity or pressure. Generally, centrifugal fans rather than axial fans are used in ventilating systems because centrifugal fans are quieter and less expensive to install and operate. Although centrifugals cope better with uncertain or fluctuating airflow conditions than axial fans, their efficiency is generally lower. Centrifugals can be divided into three classes depending on the shape and setting of the fan wheel blades. Although their applications and advantages overlap, there are distinct differences (Table 9.2):

Radial Blade Fans

Radial blade fans (Figure 9.13a) are used for dust systems since their flat radial blades tend to be self-cleaning. These fans also have large openings between blades and are therefore less likely to clog. They can be built with thick blades to withstand erosion and impact damage from airborne solids. Typical operating ranges are from small units to fans handling 100,000 ft³/min at 20 in. of water static pressure. Their major disadvantage is that they are the least efficient fan for local exhaust systems. Their heavy construction adds to their cost.

The static pressure rating curve (Figure 9.14) shows that the operating point for this fan should be selected well to the right of the peak in the pressure curve. This will avoid pulsing flow since the fan's pressure capacity varies little over a wide range of airflow volumes to the left of the peak. The horsepower curve rises in an almost straight line over the operating range of the fan.

Forward Curved Blade Fans

Forward curved blade fans (Figure 9.13b) are useful when large volumes of air must be moved against moderate pressures (0 to 5 in. of water) with low noise levels. These fans have many cup-shaped blades that accelerate the air and discharge it at a higher velocity than the fan wheel tip is moving. The shape of the fan housing converts the high air

Table 9.2 Comparison of Centrifugal Fans

Factor	Forward Curved Blade	Backward Inclined Blade	Radial Blade
First cost	Low	High	Medium
Efficiency	Low	High	Medium
Operational stability	Poor	Medium	Medium
Tip speed	Low	High	Medium
Abrasion resistance	Poor	Medium	Good
Sticky material handling	Poor	Medium	Good

Source: Reference 10.

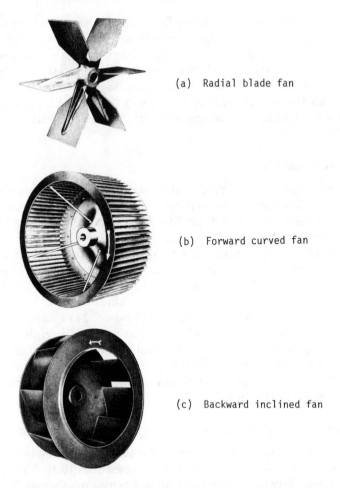

(a) Radial blade fan

(b) Forward curved fan

(c) Backward inclined fan

Figure 9.13 The shape of the fan blades for centrifugal fans. (Source: Reference 11)

velocity into static pressure. Since this is an inefficient process, the fan's overall efficiency is low. This poor efficiency limits the fan's application since some other types of fans are more efficient in higher pressure systems. The main advantage of the forward curved fan is the high blade discharge velocity, which makes possible high air speeds with relatively low fan rotating speeds. Since fan noise is related to fan speed, a low rotating speed makes the forward curved blade fan quieter, for some low and moderate pressure systems, than other fans. The high air velocity across the blades precludes its use when erosive materials are in the airstream.

The static pressure rating curve for the forward curved fan (Figure 9.15) has a valley caused by blade inefficiency at low air volumes and

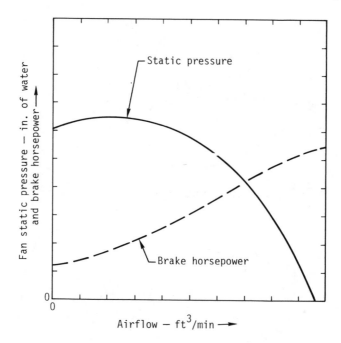

Figure 9.14   Fan curves for a typical radial blade fan.

a peak where air that follows the contour of the blade surface has the greatest velocity. Beyond the peak, the air starts to break away from the fan blade. This peak is important in fan selection if the fan is operating on the curve near the peak (Figure 9.16). In this case, even minor changes in system pressure can cause severe fluctuations in air volume through the system. This pulsing flow can occur with other centrifugal fans but is more severe with forward curved blades. To avoid pulsing, the optimum operating point should be well to the right of the peak.

Note also from Figure 9.15 that the horsepower curve rises sharply with increasing volume. If the system has less resistance (a lower static pressure) than calculated, the fan will operate at a higher flow rate through the system. As shown in Figure 9.17, a higher flow rate can result in excessive power costs, since the horsepower curve rises so sharply. This feature is another disadvantage of the forward curved fan.

Backward Inclined Blade Fans

Since backward inclined blade fans (Figure 9.13c) are more efficient than the forward curved fan, they are used more and more for handling large volumes of air with little dust. The improved efficiency occurs because the fan blades cause the pressure increase directly as the wheel rotates; the velocity of air leaving the wheel is relatively low.[11] This low blade

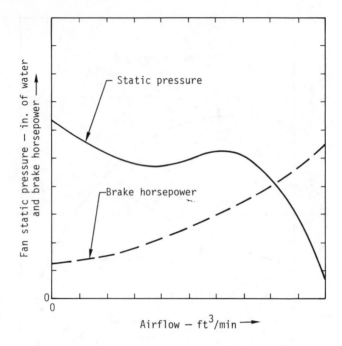

Figure 9.15   Fan curves for a typical forward curved blade fan.

discharge velocity is something of a disadvantage in larger fans, where high rotating speeds are needed to develop high air velocities. The result of these high rotating speeds is higher stresses, which necessitate heavy wheel construction and strong shafts and bearings, all of which raise initial costs.[12]

A big advantage of the backward inclined blade fan is the shape of the horsepower curve (Figure 9.18). Unlike the forward curved and radial blade fans, the horsepower curve does not rise sharply at higher volumes. Instead, the horsepower curve peaks to the right of the static pressure curve. If the actual system static pressure is less than calculated, the power costs will not rise indefinitely but will peak and decline at high airflows. This feature makes these fans a good choice for systems where initial pressure drop calculations are in doubt or for systems where pressure fluctuations occur.

Airfoil Fans

Airfoil fans are a modification of the backward inclined blade fan. The blades of airfoil fans are shaped like the cross-section of an airplane wing (Figure 9.19). This shape reduces noise and allows the fan to function smoothly without pulsing through its entire operating range. This is an

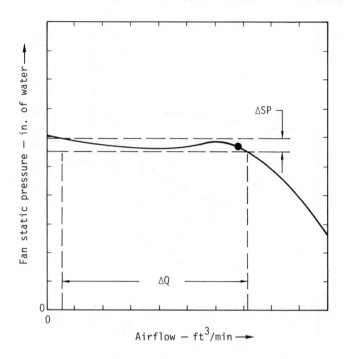

Figure 9.16 If the operating point for a forward curved blade fan is too close to the static pressure peak, wide variations in airflow will result from small variations in fan static pressure.

improvement over the backward inclined fan, which cannot operate smoothly to the left of its static pressure curve peak. The shape of the airfoil fan blades adds some structural strength to the fan wheel, thereby minimizing one problem with the backward inclined blade fan. Otherwise the airfoil fan's application is similar to that of the backward inclined fan.

Axial Fans

A screw or propeller action produces airflow in axial fans; the air travels parallel to the fan shaft and leaves the fan in the same direction as it entered. The three different types of axial fans share the advantages of compactness, low initial cost, and high mechanical efficiency. Their disadvantages include relatively high noise levels, low pressure capability (less than 10 in. of water static pressure), and their unsuitability for hot or contaminated air when the fan motor is installed inside the duct.

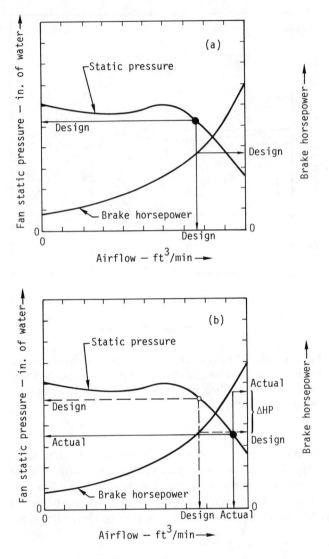

Figure 9.17   The shape of the brake horsepower curve for forward curved blade fans is important. In (a) the fan operating point and horsepower based on design calculations is shown. If the airflow through the installed system is higher than planned (b), the power consumption is much higher than expected (ΔHP) due to the rising horsepower curve.

Propeller Fans

Simple propeller fans (Figure 9.20) are not used in duct ventilation systems because they do not produce pressure (either positive pressure

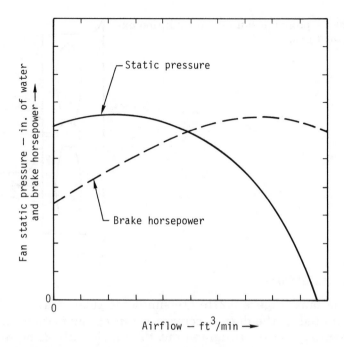

Figure 9.18 Fan curves for a typical backward inclined blade fan.

or suction). They are suitable for use as window fans or roof ventilators, where there is no real resistance to airflow.

Tube Axial Fans

Tube axial fans (Figure 9.21a) are special propeller fans mounted inside a duct. The blades are specially shaped to enable the fan to move air against low (0 to 3 in. of water static pressure) resistance.

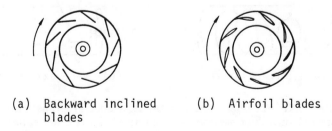

(a) Backward inclined blades

(b) Airfoil blades

Figure 9.19 The difference between backward inclined blade fans and airfoil fans is the shape of the blades.

Figure 9.20 A propeller fan used for dilution exhaust applications.

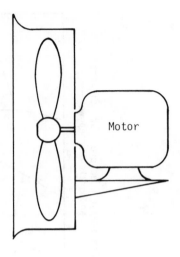

Vane Axial Fans

Vane axial fans (Figure 9.21b) are similar to tube axial fans but have vanes mounted in the duct to convert spinning air motion into higher static pressure and to straighten out the moving air. If noise is not a problem, they are useful in systems with static pressures of about 2 to 10 in. of water. Vane axial fans are available with adjustable pitch blades for systems with changing static pressure requirements. The blade pitch can be adjusted to change the fan rating curve. In this way the fan can operate at different pressure-volume relationships without sacrificing its high efficiency.[13]

Rating curves for vane axial fans vary depending on blade shape, straightening vanes, and other factors. In general both the static pressure and horsepower curves (Figure 9.22) exhibit a peak at 50 to 70%

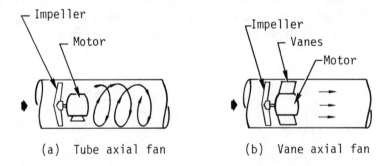

(a)   Tube axial fan   (b)   Vane axial fan

Figure 9.21 A tube axial fan (a) is a special propeller fan mounted in a duct section. A vane axial fan (b) has vanes to straighten airflow and increase the pressure generated by the fan.

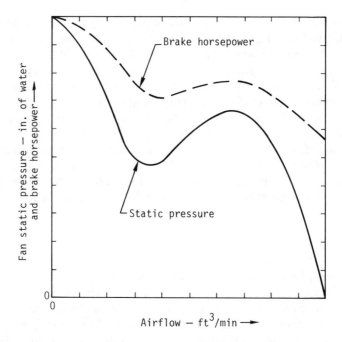

Figure 9.22 Fan curves for a vane axial fan.

of the fan's wide open (no resistance) capacity. To avoid pulsing flow in the system due to small pressure changes, the operating point is selected to the right of the pressure peak (see Figure 9.16). The horsepower curve also declines to the right of the peak, illustrating why axial fans are so efficient. To the left of the peak the fan blades become quite inefficient in moving the air as the volume drops off. This decline in efficiency is due to disturbances in airflow patterns over the blades and causes the steep rise in horsepower at lower flows. The noise level in the system usually increases as the horsepower increases.

A combination axial-centrifugal fan is also available from some manufacturers. This fan has a backward inclined or airfoil blade wheel mounted in-line in a duct section. The fan shaft is parallel to the duct. Splitter vanes and deflectors divert the air through the fan wheel, then straighten the air out downstream to convert the spinning air velocity into static pressure. The fan has a rating curve that is similar to the backward inclined blade centrifugal fan (Figure 9.18), but the axial-centrifugal unit is more compact and requires less installation space. It is quieter than pure axial fans.

Throughout this section the shape of the rating curve for different fan types has been stressed. The shape of the static pressure curve and the brake horsepower curve are important because they illustrate what will happen if the pressure drop through the system is different from that calculated during design. For example, choosing a forward curved blade fan for a system where the pressure loss could be significantly less than calculated is a mistake since the power costs rise sharply with high airflow volumes (Figure 9.17). For all fans, the shape of the static pressure curve shows the useful operating range of the fan in volume output and pressure. Although using fan rating tables rather than curves deprives you of the graphic overview of fan performance, it is more convenient, since fan manufacturers have selected the appropriate fans from their lines of different fans and motor sizes. Keep in mind that all the data points on the tables are taken from individual fan curves. Each fan is capable of a wide range of pressure-volume output combinations.

FAN NOISE

Fan noise can be a problem with some ventilation systems. Noise complaints can occur in areas that are served by the system, in locations near the fan, and occasionally in areas remote from any part of the ventilation system. The best solution to most fan noise problems is to select a quiet fan and to provide the proper mounting. If the noise source is turbulence in the air moving through the fan, attempts to quiet the fan can be expensive and ineffective.

Turbulent vs. Mechanical Noise

Fans make noise in two ways:[14]

- Turbulent noise from air moving over the fan blades, impacting the housing and changing direction at the inlet and outlet. This noise travels through the ducts and into the workrooms served by the exhaust system. The magnitude of turbulent noise from these different sources is the primary reason why some types of fans are quieter than others. Table 9.3 lists the major noise sources for forward curved and backward inclined fans. Minimize turbulent noise problems by selecting a quiet fan from fan catalogs. Table 9.3 does not include radial blade fans because they are inherently noisy.
- Mechanical noise from the fan motor bearings and drive as well as the noise radiated from the fan housing. Vibration noise from unbalanced moving parts is also a form of mechanical noise. To minimize mechanical noise problems, use flexible connections between ducts and fan and provide an inertial base or vibration isolators for mounting the fan (Figure 9.23). The area where the fan is located should be as insensitive to noise as possible. Occasionally a building struc-

Table 9.3 Significant Noise Sources for Centrifugal Fans

Noise Source	Cause	Forward Curved Blade Fan	Backward Inclined Blade Fan
Inlet	Blades cutting air	X	X
Blades	Air separating from blades	X	X
Blade outer edges	Air from top and bottom of blades merging	X	X
Housing	Airstreams changing speed and direction	X	

Source: Reference 14.

ture is too weak or too flexible to withstand the motion of the spinning fan, or live load, without excessive vibration. This potential problem should be identified before fan location is decided.

Turbulent Noise Solutions

Prevent turbulent noise problems by selecting a quiet fan from fan manufacturers' catalogs. The best selections for quiet operations are indicated on fan rating tables or diagrams (Figure 9.6). Since the zone of quiet operation usually matches the maximum operating efficiency range (Figure 9.24), this is a good way to select fans. Although this procedure helps you pick the quietest fan of that make and model, it does not permit comparison of different makes of fans or even different types of fans (for

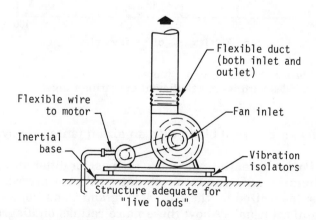

Figure 9.23 Illustration of ways to reduce noise and vibration transmitted by the fan to the rest of the building.

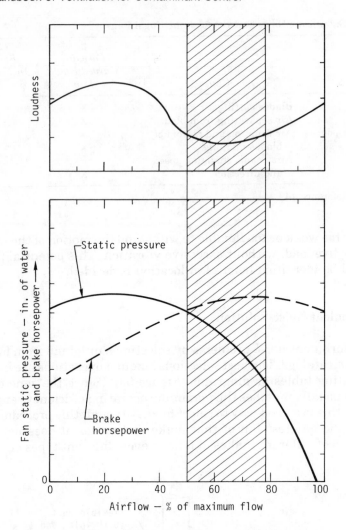

Figure 9.24   The zone of quietest operation for a fan coincides with its most efficient operating range.

example, a forward curved blade with an airfoil fan) made by the same company.

Sound Power Level Ratings (noise ratings) for different fans can be calculated from manufacturers' noise data collected according to test procedures standardized by the Air Moving and Conditioning Association. Fan catalogs usually show these noise ratings on diagrams or illustrate the calculations needed to find a rating for comparison with ratings of other fans.

The quietest type of fan is the airfoil backward inclined blade fan followed by the plain backward inclined blade fan. The noisiest fans are the radial blade and axial flow fans. The forward curved blade fan is also noisier than the backward inclined fan except that in small sizes (generally about 18 in. wheel diameter or less) at low static pressure the forward curved blade fan is quieter than a similar size backward inclined fan. This is because the rotating speed of the backward inclined fan is much greater. A 15-in. diameter forward curved blade fan operating against 1.0 in. of water static pressure will pull over 2000 ft³/min. This is enough for many small ventilation systems. In general, however, only airfoil fans should be considered for large installations where noise is a problem unless the airborne solids in the exhaust stream make the radial blade fan a necessity.

Here are some additional guidelines for fan noise control.

- In areas where noise is critical, reduce the system resistance. Since fan noise, static pressure, and resistance are all proportional to each other, reducing system resistance will reduce noise directly.
- When all other factors are equal, use larger, slower fans rather than smaller, higher speed fans.
- Pulsation noise can result from poor fan inlet conditions such as inlet boxes (Figure 9.8) that load only one side of the fan wheel with air. These poor connections should be eliminated as part of the system design procedure.
- Some older fan design specifications may try to control noise by putting a ceiling on fan outlet velocities. This concept is no longer valid; fan noise depends on fan efficiency, not on fan outlet velocity.
- For supercritical noise applications, the fan motor, drive, and bearing noise may have to be considered. If the isolation methods in Figure 9.23 are used, however, these mechanical noises probably will not be a problem compared to turbulent noise.

FAN LAWS GOVERN OPERATION

There is a series of statements that explain how fans work. They are called Fan Laws and are used to construct the fan rating tables and curves. They can also be used to help you decide how to modify a fan's operation so that it works properly in your system. These laws could have been defined earlier in the chapter, but they also make a good summary of fan selection principles.

Three Key Laws

The three most descriptive laws describe the relationship of volume, fan static pressure, and brake horsepower to fan speed:

- Changes in volume (ft³/min) vary directly with changes in fan speed. For a given fan doubling the speed will double the volume output.

$$\frac{R_1}{R_2} = \frac{Q_1}{Q_2} \qquad (9.3)$$

where R = fan rotating speed, rev/min
   Q = airflow, ft³/min

- Changes in static pressure vary directly with the square of changes in fan speed. If you double the fan speed, the static pressure generated by the fan increases by a factor of four. A corollary to this law is that static pressure also varies directly with the square of changes in fan volume.

$$\frac{FSP_1}{FSP_2} = \left(\frac{R_1}{R_2}\right)^2 \qquad (9.4)$$

where FSP = fan static pressure, inches of water
   and from Equations 9.3 and 9.4,

$$\frac{FSP_1}{FSP_2} = \left(\frac{Q_1}{Q_2}\right)^2 \qquad (9.5)$$

- Changes in brake horsepower vary directly with the cube of changes in fan speed. Doubling the speed of a fan increases brake horsepower by a factor of 8 (2 × 2 × 2 = 8).

$$\frac{BHP_1}{BHP_2} = \left(\frac{R_1}{R_2}\right)^3 \qquad (9.6)$$

where BHP = Brake Horsepower

Since all three of these fan laws act together, any change in fan speed to increase volume output also increases fan static pressure and brake horsepower. Especially as power costs increase due to rising electric rates, the jump in brake horsepower should be considered before increasing fan speed. If a fan is too small to do the job economically, it may be less expensive in the long run to replace it with a larger fan that uses less power than that consumed in speeding up the small fan.

Example: In a manufacturer's catalog a fan is rated as delivering 10,500 ft³/min of air at 3 in. of water fan static pressure when running at 400 rev/min and requiring 6.2 horsepower. If the fan speed is increased to 500 rev/min, determine the volume, static pressure, and horsepower. Assume standard conditions.

Answer: Capacity: $Q = 10,500 \left(\dfrac{500}{400}\right) = 13,125$ ft³/min

Static Pressure: $FSP = 3 \left(\dfrac{500}{400}\right)^2 = 4.7$ in. of water

Horsepower: $HP = 6.2 \left(\dfrac{500}{400}\right)^3 = 12.1$ horsepower

There are additional fan laws besides the basic three. They deal with the effect of different fan size and air density on fan performance. They are summarized in Table 9.4.

IS THE FAN WORKING PROPERLY?

Since the fan is the only moving part in most ventilating systems, it often receives a lot of scrutiny when the system is not working properly. This chapter has focused on the different types of fans and their selection for ventilation systems. If you have a fan in an existing system and the system is not working properly, you have a different problem. Chapter 12 deals with testing all ventilation system components, including fans. Chapter 13 tells you how to use test data in reviewing system perfor-

Table 9.4 Fan Laws

	Variables[a]		
Fan Speed	Fan Size	Air Density	Effect
Varies	Constant	Constant	Volume varies as fan speed
			Pressure varies as square of fan speed
			Power varies as cube of fan speed
Constant	Varies[b]	Constant	Volume varies as cube of wheel diameter
			Pressure varies as square of wheel diameter
			Tip speed varies as wheel diameter
			Power varies as fifth power of wheel diameter
Constant	Constant	Varies	Volume constant
			Pressure varies as density
			Power varies as density

[a]Assumes constant system (hoods, and duct lengths and diameters do not change).
[b]Assumes constant fan proportions as when selecting different wheel size of same fan type.

mance to correct ventilation system problems. Chapter 11 offers guidelines for designing a system that will do the job at minimum cost.

SUMMARY

The fan generates the suction that draws contaminated air into the hoods and through the ducts and air cleaner. Fan selection involves choosing the proper fan to match the airflow and static pressure requirements of the system. In addition, it is important that the fan exhibit stable performance at its operating point. This means that small variations in static pressure (resistance to airflow) in the system do not cause large fluctuations in airflow.

Data on fan performance are listed in fan rating tables in manufacturers' catalogs. These tables show airflows at different static pressures for each fan rotating speed. The relationship among the three main variables—airflow, static pressure, and rotating speed—as well as power consumption can be calculated from the Fan Laws once their values at one set of operating conditions are known.

REFERENCES

1. AMCA Bulletin 110. "Standards, Definitions, Terms, and Test Codes for Centrifugal, Axial and Propeller Fans" (Park Ridge, Illinois: Air Moving and Air Conditioning Association, Inc., 1952).
2. Catalog, New York Blower Company, Chicago, Illinois.
3. Catalog, Chicago Blower Corporation, Glendale Heights, Illinois.
4. ACGIH Committee on Industrial Ventilation. *Industrial Ventilation—A Manual of Recommended Practice,* 17th Ed. (Lansing, Michigan: American Conference of Governmental Industrial Hygienists, 1982).
5. Geissler, H. "Purchased Fan Performance," Reprint No. 5483 (Pittsburgh, Pennsylvania: Westinghouse Electric Corporation, 1959).
6. Trickler, C.J. "Field Testing of Fan Systems," Engineering Letter No. E-3 (Chicago, Illinois: The N.Y. Blower Company).
7. Trickler, C.J. "Effect of System Design on the Fan," Engineering Letter No. E-4 (Chicago, Illinois: The N.Y. Blower Company).
8. Tracy, W.E. "Fan Connections," Reprint No. 5100 (Pittsburgh, Pennsylvania: Westinghouse Electric Corporation, 1955).
9. National Institute for Occupational Safety and Health. *The Industrial Environment—Its Evaluation and Control* (Washington, D.C.: U.S. Government Printing Office, 1973).
10. Cheremisinoff, P.N. and R.A. Young. "Fans and Blowers," *Pollution Engineering* 6, No. 7 (1974).
11. Trickler, C.J. "Fundamental Characteristics of Centrifugal Fans," Engineering Letter No. E-1 (Chicago, Illinois: The N.Y. Blower Company).

12. Rogers, A.N. "Selection of Fan Types," Reprint No. 5312 (Pittsburgh, Pennsylvania: Westinghouse Electric Corporation, 1957).
13. American National Standard Z 9.2-1971. "Fundamentals Governing the Design and Operation of Local Exhaust Systems," (New York, New York: American National Standards Institute, 1972).
14. Trickler, C.J. "How to Select Centrifugal Fans for Quiet Operation," Engineering Letter No. E-13 (Chicago, Illinois: The N.Y. Blower Company).

CHAPTER 10

Ventilation for High-Toxicity or High-Nuisance Contaminants

Most of the hood selection and system design information presented so far applies to standard industrial operations such as welding, metal working, or open surface tanks using low or moderate toxicity materials. Hood selection and airflow calculations are based on capturing or containing enough of the contaminants to reduce employee exposures to acceptable levels. Often 100% control is neither necessary nor achievable with standard ventilation systems.

The adequacy of a ventilation system is determined by evaluating employee exposures with the system in operation. If the exposures are within acceptable limits compared to OSHA permissible exposure standards, Threshold Limit Values (TLVs), or other toxicological guidelines, the system is providing sufficient protection to the workers. For high-toxicity or high-nuisance contaminants the allowable concentration may be so low that hoods designed using standard design criteria permit too much of the contaminant to escape into the workroom. Special emphasis is required to reduce the amount of contaminants to acceptable levels.

As an illustration consider the raw cotton feed hopper hood installed in a cotton yarn manufacturing plant (Figure 10.1). Raw cotton is dumped into the feed hopper and is then carried to cleaning, picking, carding, and other steps used to draw cotton yarn. The hood is designed to pull an average of 75 ft/min through the open area[1] to keep cotton dust levels below 0.2 mg/m³ for this material. But would this same hood be suitable for a feed hopper handling asbestos fiber rather than raw cotton? Probably not, since asbestos is recognized as being more toxic than cotton dust and has a very low exposure limit. Figure 10.2 illustrates an as-

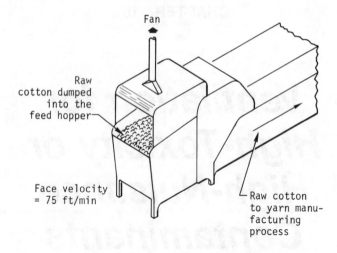

Figure 10.1   Low face velocity provides adequate control of lower-toxicity materials such as cotton dust.

bestos fiber bag opening and dumping station used in a plant manufacturing asbestos pipe.[2] Sealed bags of asbestos fiber are placed in the hood and opened; the fiber is then dumped into a hopper. A ventilated waste bag is provided for disposal of empty fiber bags. With a face velocity of 200 ft/min, airborne levels of asbestos fibers are within acceptable limits.

Figure 10.2   A face velocity of 200 ft/min is needed to control the release of asbestos at this station where bags of asbestos fiber are opened and dumped. The ventilated plastic bag on the right holds empty asbestos fiber bags for disposal.

ZERO EXPOSURE STANDARDS

In addition to substances with low exposure standards, some carcinogens have essentially zero allowable exposure levels.The ACGIH Threshold Limit Value Committee has identified human carcinogenic substances associated with industrial processes. For some of these substances they recommend that:[3]

> . . . no exposure or contact by any route, respiratory, skin or oral, as detected by the most sensitive methods, shall be permitted. The worker should be properly equipped to insure virtually no contact with the carcinogen.

The substances falling within this classification can be found in the current TLV listing.[3]

Federal OSHA Standards also contain provisions to reduce exposures to some carcinogens to near zero levels by restricting their use to isolated systems or requiring ventilation.[4] The standards define isolated, closed, and open systems (Table 10.1) according to the amount of protection they afford against release of the substance. Operations in open vessel systems are prohibited and ventilation is required when a closed system is opened for material loading or other reasons. Personal protective equipment, washing and showering requirements, and other provisions are designed to reduce exposures to as close to zero as possible.

HIGH-NUISANCE MATERIALS

High-nuisance materials is a nebulous term describing contaminants that cause nuisance problems in very low concentrations. Odor is the most common problem since physiological effects such as eye irritation or throat irritation would be reflected in the toxicity rating and the permissible exposure to the substance. High-nuisance materials are in-

Table 10.1 OSHA Definitions of Systems for Certain Carcinogens

Isolated System	Fully enclosed structure other than the vessel of containment which is impervious to the passage of the material and would prevent escape due to leakage or spillage from the vessel of containment. A glove box enclosure is an example.
Closed System	Operation where containment prevents the release of the material. Sealed containers and closed piping systems are examples.
Open Vessel System	Operation in an open vessel which is not an isolated system, a laboratory hood nor any other system providing equivalent protection.

Source: Reference 4.

cluded in this chapter since the control techniques for high-toxicity contaminants also apply to nuisance abatement. If the material has a strong
odor or other nuisance factor in very low concentrations, standard ventilation system design criteria may allow too much of the material to
escape into the workroom or community environment. The special techniques discussed in this chapter may help solve the problem if ventilation, rather than another control method, is selected. To avoid repetition
throughout the remainder of this chapter, high-toxicity or high-nuisance
materials will be referred to as high-toxicity materials, but the guidelines apply to high-nuisance materials as well.

HOOD DESIGN

The hood is the best place to start designing a ventilation system for
high-toxicity materials. If the contaminants escape from the hood into
the workroom, the ventilation system will not provide the intended protection; therefore, the hood should be the most important part of the
system. There are three types of hoods: capturing hoods, enclosures, and
receiving hoods (Figure 6.7). For high-toxicity materials the receiving
hood is usually unacceptable since contaminant control is insufficient
with this hood type. In fact, receiving hoods are rarely used for protection
against inhalation hazards when one of the other two hood types can be
used. The design criteria for capturing hoods and enclosures can often
be modified for high-toxicity systems.

Capturing Hoods

Capturing hoods control contaminants by reaching outside the hood to
draw them in (Figure 6.17). A NIOSH-sponsored study[5] of ventilation
for grinding, buffing, and polishing hoods showed that grinder hoods
(Figure 10.3) with recommended airflow do an adequate job of controlling
contaminants for workpieces, abrasives, and abrasive supports made of
inert materials (meaning biologically inert when inhaled) with a current
TLV of 10 mg/m³. However, the study found that for materials with a
TLV below 10 mg/m³ the conventional grinder hood was no longer adequate for long exposures because of escaping fine particles entrained
with the spinning grinding wheel (Figure 10.4). For grinding, polishing,
or buffing materials, such as copper, chromium, or nickel, with exposure
standards between 1 and 10 mg/m³, either an auxiliary hood (Figure
10.4b) along with a conventional hood or a ventilated enclosure is needed.
Finally, for materials with an allowable exposure level below 1 mg/m³
like beryllium, lead, or silver, or in cases when sand castings are ground
with a free silica hazard from the sand, an enclosing hood and respiratory

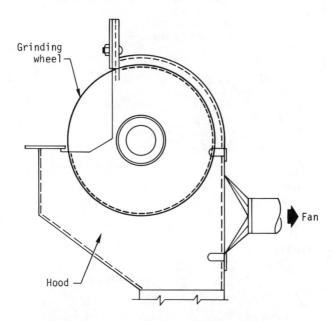

Figure 10.3 The conventional pedestal grinder hood design does not control enough contaminants to be used with highly toxic materials.

protection for those entering the enclosure are needed. Table 10.2 summarizes the recommendations of this study. Remember that the hood on a grinder or cut-off wheel provides protection if the wheel shatters while spinning.[6] Any enclosure design should provide suitable structural strength if the conventional grinder hood is not used.

The OSHA ventilation standards for open surface tanks[7] require different control velocities depending on the toxicity of the contaminants. The *Hazard Potential* is an A through D rating for materials based on their toxicity and flash point (Table 10.3). The Hazard Potential, along with the rate of contaminant evolution, determines how much control velocity is required. For materials in the highest toxicity category (A), the ventilation control velocity requirements are about double those for low-toxicity (D) rated materials.

Both the ACGIH *Industrial Ventilation Manual*[8] and the ANSI Z 9.2-1971 standard for local ventilation[9] contain capture velocity recommendations (Table 6.5). Although the toxicity of the material must be considered when selecting a capture velocity at the high or low end of each range in Table 6.5, the primary factor governing the control velocity is how much energy or velocity the contaminants have when released from the process.

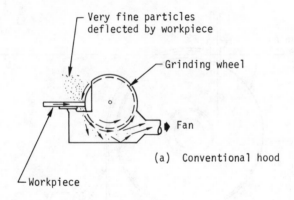

(a) Conventional hood

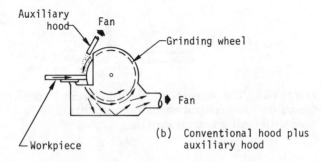

(b) Conventional hood plus
     auxiliary hood

Figure 10.4   An auxiliary hood captures the very fine
particles that escape from the conventional grinder
hood.

Here is a summary of things to consider when using capturing hoods
with highly toxic contaminants:

- For some applications capturing hoods will never capture enough
  contaminants to be usable with high-toxicity materials. Conventional
  grinding hoods are an example.
- In many cases capturing hoods will control the contaminants if the
  capture velocity is increased sufficiently. For example, a flanged suction opening (Figure 10.5) that develops a capture velocity of 1000–
  2000 ft/min about 2 in. from the opening can be used as a close capture
  hood to control airborne contaminants when milling or drilling toxic
  metals.
- Crossdrafts have a severe impact on the efficiency of capturing hoods
  (Figure 6.15). Choose the hood location carefully to avoid crossdrafts.

Table 10.2 Ventilation for Grinding, Buffing, and Polishing

Category	TLV of Contaminant	Typical Operations	Ventilation Requirements
I	10 mg/m³	All work with "inert" workpieces, abrasives, and abrasive supports.	Standard ventilation system
II	1–10 mg/m³	GBP[a] copper, chromium, or nickel	Auxiliary hood combined with standard hood or good enclosures
III	Less than 1 mg/m³	GBP[a] lead, cobalt, silver, or beryllium Grinding sand castings Using grinding wheels containing lead or silica Buffing with silica abrasive	Total enclosure and respiratory protection for personnel entering the enclosure

[a]GBP: Grinding, buffing, and polishing.
Source: Reference 5.

Table 10.3   OSHA Hazard Potential Rating for Open Surface Tanks

	Permissible Exposure Limit for Material		
Hazard Potential	Gas or Vapor (ppm)	Mist, (mg/m³)	Flash Point (°F)
A	0–10	0–0.1	—
B	11–100	0.11–1.0	Under 100
C	101–500	1.1–10	100–200
D	Over 500	Over 10	Over 200

Source: Reference 7.

Enclosures

Enclosures are often the best choice for highly toxic materials. These hoods surround the contaminant source and function by preventing the escape of material released inside the hood. Enclosures have these advantages, especially for high-toxicity applications:

- Essentially complete control of contaminants can be achieved with the proper enclosure design.
- The volumetric airflow to control contaminants is usually less than for capturing hoods.
- Enclosures are less affected by crossdrafts and other factors that reduce capturing hood efficiency.

Designing Enclosures

The ACGIH *Manual* considers toxicity and nuisance potential in the design recommendations for laboratory hoods (Table 10.4). For high-toxicity substances an average inward velocity of 150 ft/min is recommended; for low toxicity substances 100 ft/min is adequate. The federal OSHA standards for some cancer-suspect agents[10] require an average

Table 10.4   Recommended Face Velocity for Laboratory Hoods

Material Used in Hood	Average Face Velocity (ft/min)
Nuisance or corrosive	100
Moderately toxic	100
Tracer quantities of radioisotopes	100
Highly toxic (TLV ≤ 10 ppm or 0.1 mg/m³)	150
Low maximum permissible concentration radioisotope	150

Source: Reference 8.

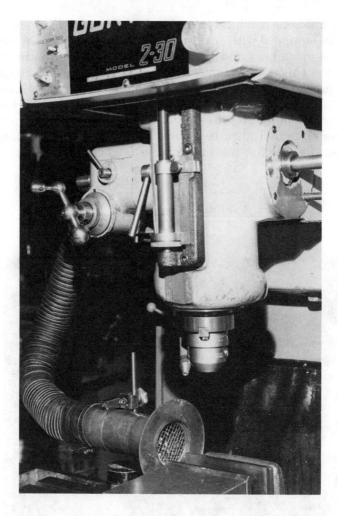

Figure 10.5 The close capture hood on this milling machine generates 1000–2000 ft/min velocity about 2 inches from the hood opening.

inward face velocity of at least 150 ft/min with a minimum velocity of 125 ft/min at any point in the door opening. However, some studies indicate that excessively high face velocities can cause turbulence, which can increase the hood user's exposures to chemicals handled inside the laboratory hood as discussed in Chapter 6.

The important design parameter for enclosures is the inward face velocity through openings in the enclosure. Since volume equals velocity multiplied by open area, reducing the area of openings to a minimum also reduces the volumetric airflow requirement.

Many different machine tools, other equipment, and industrial processes can be equipped with enclosures. However, enclosures often hinder access to the machinery and may limit the size of objects worked on the machines. For this reason, removable enclosures or units with doors can be used to allow nontoxic work on the machinery after decontamination. Enclosures for machine tools such as lathes and milling machines (Figure 10.6) are designed to provide 300 ft/min air velocity through openings.[8] The lathe in Figure 10.6 has a sliding door that is almost closed during operation. If the total open area is about 50 in.², the required airflow is:

$$300 \ \frac{\text{ft}}{\text{min}} \times 50 \ \text{in.}^2 \times \frac{\text{ft}^2}{144 \ \text{in.}^2} = 104 \ \text{ft}^3/\text{min}$$

Figure 10.6 Enclosure hood on a lathe used for machining highly toxic materials.

This is a modest airflow compared to the air volume needed to exhaust the lathe with a capturing-type hood.

Glove Box Enclosures

The glove, or dry, box (Figure 10.7) is the ultimate enclosure. It is a sealed or low leakage box fitted with flexible airtight gloves for manual operations inside the box. The enclosure is exhausted at a low rate, and the fan also keeps the box interior at a slightly negative pressure to prevent outward leakage. Too much negative pressure inside the box makes the gloves hard to manipulate; too little suction may allow pressurization of the box when the gloves are inserted suddenly. Pressurizing the box can allow contaminants to escape through cracks.

The airflow for glove boxes is calculated to remove contaminants and heat generated inside the enclosure, maintain the box interior at negative pressure, and prevent contaminant escape if a glove fails:

- Normal airflow is 2–15 ft³/min per 20 ft³ of box volume. The size of the air inlet into the box is selected to give this airflow.
- In addition to normal airflow the fan must be capable of moving sufficient air to develop 100 ft/min velocity through an open glove

Figure 10.7 Glove box enclosure around a small lapping machine provides total containment at low airflow.

port in case a glove pulls loose from its retaining rings. Depending on the diameter of the glove port, the emergency airflow is usually 35–40 ft³/min.[11]

• If combustible gases, vapors, or aerosols are released inside the box, the normal airflow must be sufficient to dilute the concentration well below the lower explosive limit (LEL). Using Equation 1.2 to calculate the flow rate:

$$Q = \frac{6.7 \ (sp \ gr) \ (W) \ (100) \ (C)}{(M) \ (LEL) \ (B)} \qquad (10.1)$$

where Q = dilution airflow, ft³/min
  sp gr = specific gravity of liquid
    M = molecular weight
    W = amount of combustible material used, pints/hr
  LEL = lower explosive limit, percent
    C = safety factor representing how much excess airflow
        is needed to reduce the concentration below the
        LEL; C = 4 if 25% of the LEL is acceptable,
        C = 10 for 10% of the LEL, and so forth
    B = correction for elevated temperatures since the LEL
        is lowered. B = 1 for temperatures up to 250°F;
        B = 0.7 for temperatures above 250°F.

Example: Toluene is used in a glove box at the rate of 0.5 pt/hr. Assuming the temperature is 70°F, how much airflow is needed to reduce the concentration to 10% of the LEL?

Answer: The molecular weight of toluene is 92.13, the specific gravity is 0.9 and the LEL is 1.27%.[12]

$$Q = \frac{(6.7) \ (0.9) \ (0.5 \ pt/hr) \ (100) \ (10)}{(92.13) \ (1.27) \ (1.0)}$$
$$Q = 25.8 \ ft^3/min$$

• Heat build-up inside the box can be a problem if hot plates or other heat sources are used in the box. A temperature increase that does not exceed 15°F above the ambient room temperature is usually acceptable. Above this value the worker's arms and hands may perspire excessively inside the gloves. The ventilation flow rate should be increased to remove the excess heat; however, if the calculated rate is more than one air change per minute, either auxiliary cooling or a heat shield should be considered. The airflow to dilute the heat load from sensible heat (the heat radiated by the hot object) is calculated from this equation:[11]

$$Q = \frac{H}{1.08\ (\Delta t)}$$ (10.2)

where Q = dilution airflow, ft³/min
    H = total sensible heat load, Btu/hr
    Δt = desired temperature increase in box relative to
        ambient air temperature, °F

Example: A 500-watt hot plate is used inside a glove box with a volume of 15 ft³. What airflow is needed to limit temperature increase in the box to 15°F above room air?

Answer: One watt is equivalent to 3.42 Btu/hr[13] so a 500-watt hot plate produced 1710 Btu/hr assuming all of the electrical energy is converted to heat. From Equation 10.2:

$$Q = \frac{1710\ \text{Btu/hr}}{1.08\ (15°F)} = 105.5\ \text{ft}^3/\text{min}$$

Since this airflow is almost 10 times the volume of the box, a heat shield or other heat control technique should be investigated to reduce airflow.

Airflow rate in a glove box is governed by three critical parameters. The highest rates are required in cases where heat or flammable vapors must be removed. If these factors are not involved, then the minimum airflow that will keep the box under negative pressure is acceptable. Regardless of the normal airflow, the fan must be capable of generating the required emergency airflow if a glove fails. If the normal airflow is low compared to the emergency airflow, the latter will probably govern the fan size needed for the glove box exhaust system. A damper in the inlet may be needed to adjust airflow to proper values during normal operations.

Some glove boxes are filled with inert atmospheres, such as nitrogen or argon. These gases exclude oxygen, moisture, or other components of room air that cause interference with operations. In order to lower the leakage rate and inert gas consumption to an acceptable level, the design and construction criteria for inert enclosures are more stringent than are those for air-exhausted boxes.

If a glove box is used as an isolated system to prevent exposures to certain carcinogens, the design should include an air lock arrangement. In such an arrangement material can be transferred into and out of the box while preventing escape of contaminants inside the box. An air lock is also required for boxes with inert or other special atmospheres to keep air and moisture out of the box (Figure 10.8). For materials requiring complete isolation, Figure 10.9 illustrates how materials can be transferred into and out of a glove box with no exposure to the outside environment.

An important factor in determining the size of a glove box and the number and location of glove ports is the limitations of the human body

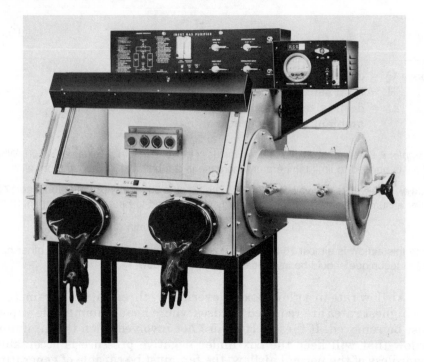

Figure 10.8   A glove box supplied with an inert atmosphere to exclude the oxygen and moisture in room air. The air lock at the right keeps air out of the box during material transfers. *(Courtesy Kewaunee Scientific Equipment Corp.)*

in reaching and manipulating. As a rule-of-thumb, work should involve reaching no more than 22 in. inside the box.[11]

AIR CLEANER

Ventilation systems for high-toxicity or high-nuisance materials almost always require an air cleaning device to remove the contaminants collected in the system before the air is discharged into the community environment. Air cleaners are selected for their efficiency in removing the contaminants in the air. Refer to Chapter 7 for different types of air cleaners. In this section special consideration is given to one type of air cleaner used to collect highly toxic particulates.

HEPA or Absolute Filters

The most efficient air cleaner for particulates is the High-Efficiency Particulate Air (HEPA) filter, also called an absolute filter (Figure 10.10).

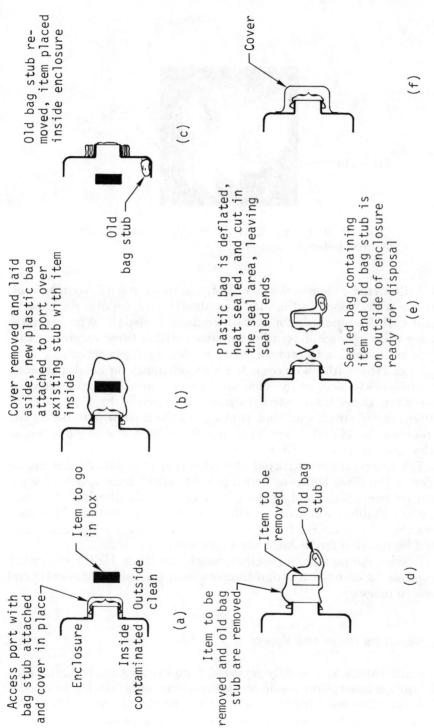

Access port with
bag stub attached
and cover in place

Item to go
in box

Enclosure
Inside
contaminated
Outside
clean

(a)

Cover removed and laid
aside, new plastic bag
attached to port over
existing stub with item
inside

(b)

Old bag stub re-
moved, item placed
inside enclosure

Old
bag stub

(c)

Item to be
removed and old bag
stub are removed

Item to be
removed
Old bag
stub

(d)

Plastic bag is deflated,
heat sealed, and cut in
the seal area, leaving
sealed ends

(e)

Sealed bag containing
item and old bag stub is
on outside of enclosure
ready for disposal

Cover

(f)

Figure 10.9 Step-by-step procedure for transferring items into and out of an isolated system. (Source: Reference 11)

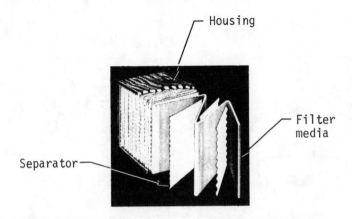

Figure 10.10   Construction of a HEPA, or absolute, filter.
(Source: Reference 14)

HEPA filters are at least 99.97% efficient in collecting a 0.3-μm aerosol. They are widely used in the nuclear industry and are also required by federal OSHA standards for certain contaminants.[10] When respirators are used instead of ventilation, small HEPA filter respirator cartridges are available to provide the same degree of protection.

Since HEPA filters have such a high efficiency for small particles, they are quickly clogged by room air dust or coarse contaminants. For this reason, these large particles must be removed by a prefilter or another type of air cleaner that is placed in the airflow path before the air reaches the HEPA filter. Filtering the inlet air to a glove box is another way to extend HEPA filter service life.

The filters are constructed of pleated paper or other filter media bonded to the filter frame or housing. Since small holes or slight separation between filter and housing will reduce the filtering efficiency below acceptable standards, careful handling is important. Manufacturers should test HEPA filters before shipping; after installation they should be retested for leakage before use.[15]

Besides damage from shock or rough handling, HEPA filters are susceptible to damage by high temperature, corrosive chemicals, and excessive moisture.

Removing Gases and Vapors

Gases and vapors are usually removed from exhaust air by using activated carbon adsorption, liquid scrubbers, or incineration. For some odor problems adding masking or counteracting chemicals may also help. See Chapter 7 for more information on gas and vapor removal.

Multiple Air Cleaners

Regardless of the type of air cleaner used, the toxicity of the material, its volatility, and the allowable exposure limits may require multiple air cleaners arranged in series, in parallel, or in both series and parallel.

Different Air Cleaners In Series

When two different air cleaners are installed in series (Figure 10.11), the first collector removes the bulk of the contaminants and prevents overloading of the final cleaner.

Identical Air Cleaners in Series

For some carcinogens and radioactive materials two identical high-efficiency air cleaners in series (Figure 10.12) may be needed. If one cleaner fails or becomes saturated, the other cleaner is available. If HEPA filters are used, two filters in series do improve overall removal efficiency,[16] but the amount of material passing through the first filter is so small that the enhanced collection is often not noticeable. The real value of identical air cleaners in series is the backup protection in case one cleaner is not operating properly. If the exhaust gas contains corrosive vapors or other substances that contribute to early air cleaner failure, then both air cleaners will probably fail and contaminants will be released. Therefore both of the air cleaners must be kept in serviceable condition and tested periodically for collection efficiency.

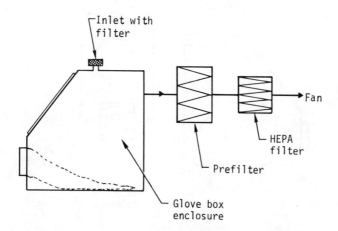

Figure 10.11   Use of a prefilter protects the HEPA filter from plugging by coarse particles.

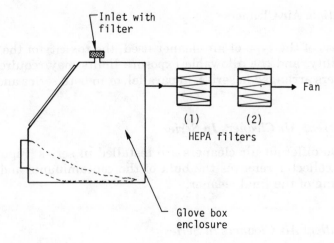

Figure 10.12   Two identical air cleaners in series provide protection against release if one air cleaner fails.

Parallel Air Cleaners

If continuous exhausting through the system is needed even while an air cleaner is being serviced or replaced, then parallel air cleaners (Figure 10.13) can be used. Proper valving allows one air cleaner to be serviced while air passes through the other unit.

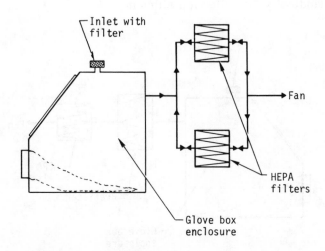

Figure 10.13   Two identical air cleaners in parallel permit one cleaner to be serviced while the exhaust system is still operating.

SYSTEM DESIGN FEATURES

After hoods or enclosures and air cleaners for the high-toxicity contaminants have been selected, the remainder of the system can be designed. Duct design and fan sizing follow Chapters 8 and 9 with a few special precautions for high-toxicity systems. Operating interlocks and contaminant or airflow sensors to monitor system performance may also be needed.[9]

Fans and Ducts

Fans should be located as close to the discharge point as possible and preferably outside of the working area. Since all ducts before the fan are at negative pressure, leaks will allow air to enter the system. Ducts on the discharge side of the fan are at positive pressure, and leaks permit air to escape from the system. For this reason, the air cleaner should be located on the suction side of the fan. If continuous airflow is needed, two fans should be installed in parallel to permit one fan to be serviced while some airflow is maintained through the system.

Ducts should be constructed so they are easy to clean, especially the ducts before the air cleaner. Round ducts are stronger than rectangular ducts and also minimize dust settling in ducts.

Operating Interlocks

Interlocks may be needed between the ventilation system and process machinery in order for the machinery to operate only when the airflow is adequate. The simplest interlock operates both the ventilating fan and machinery with the same switch. This arrangement assures that the fan is running but does not guarantee that there is adequate airflow through the system. Pressure-activated switches used as interlocks will permit the machinery to operate only when there is sufficient hood static pressure in the duct near the hood. (Hood static pressure as an indicator of airflow is discussed in Chapters 5 and 12.) If scrubbers, incinerators, electrostatic precipitators, or other air cleaners with operating controls are used, they should also be included with the fan motor in an interlock system.

Instrumentation

Instrumentation can either measure the level of air contaminants or physical characteristics of the ventilation system, such as pressure or air velocity. Either type of sensor can be connected to alarms that warn of malfunctions or to recorders that make a record of operating conditions.

Contaminant Sensors

Sensors that measure the concentration of contaminants in the work area or in the discharge stack are the best means of gauging system performance. Sensors to warn of combustible gas mixtures are probably the most common type in use. Sensors located in the workroom monitor the efficiency of hoods in controlling contaminants; sensors in the exhaust stack measure the concentration being released after the air cleaners. Sensors are not available for many contaminants. Periodic calibration of the sensing system is required.

It is important to remember that the sensors measure the airborne levels, not the workers' exposure. Each worker's exposure depends on his work pattern and how much time he spends in different concentration levels. Personal samples collected in the worker's breathing zone are needed to determine actual exposures.

Velocity and Pressure Indicators

Devices that measure important physical parameters in the system can be used in place of or along with contaminant sensors. Measuring the static pressure and air velocity at key locations gives a good indication of the system performance. In some ventilation systems measuring the temperature, relative humidity, or other factors may also be important. Although these sensors can be connected to alarms to warn of potentially unsafe conditions, it is often sufficient to locate the indicator device where workers can see it and mark safe and unsafe ranges on the meter or manometer. The glove box enclosure in Figure 10.7 is equipped with pressure sensors. A limitation of using velocity and pressure indicators in place of contaminant sensors is that some conditions appear normal when the air cleaner is not performing properly. This is covered in Chapter 7.

SUMMARY

This chapter discusses some of the special precautions to observe when using ventilation to control high-toxicity or-nuisance materials. High-toxicity substances are those that present a health risk at very low airborne levels. High-nuisance substances cause odor or other similar problems at very low concentrations in the plant or community.

Since standard ventilation system designs are usually based on controlling most of the contaminants to keep employee exposures below acceptable limits, modifications to standard designs for high-toxicity or high-nuisance materials are needed.

Possible modifications to standard designs include increasing the capture velocity for capturing hoods or the face velocity into enclosures. Complete isolation in a glove box or similar enclosure minimizes exposure without excessive airflow rates. High-efficiency air cleaners and operating interlocks and sensors to monitor system performance are usually needed.

REFERENCES

1. Barr, H.S., R.H. Hocutt, and J.B. Smith. "Cotton Dust Controls in Yarn Manufacturing," U.S. Department of Health, Education and Welfare, Publication No. (NIOSH) 74-114 (Washington, D.C.: U.S. Government Printing Office, 1974).
2. Goldfield, J. and F.E. Brandt. "Dust Control Techniques in the Asbestos Industry," *Amer. Industrial Hygiene Assoc. J.* 35, No. 12, 799 (1974).
3. American Conference of Governmental Industrial Hygienists. "Threshold Limit Values for Chemical Substances in the Workroom Environment" (Cincinnati, Ohio: ACGIH, 1983).
4. OSHA General Industry Safety and Health Regulations, U.S. Code of Federal Regulations, Title 29, Chapter XVII, Part 1910.1003 (1975).
5. Bastress, E.K., J.M. Niedzwecki, and A.E. Nugent. "Ventilation Requirements for Grinding, Buffing and Polishing Operations," U.S. Department of Health, Education and Welfare, Publication No. (NIOSH) 75-107 (Washington, D.C.: U.S. Government Printing Office, 1975).
6. American National Standard B 7.1-1970. "Safety Code for the Use, Care and Protection of Abrasive Wheels" (New York, New York: American National Standards Institute, 1970).
7. "OSHA General Industry Safety and Health Regulations," U.S. Code of Federal Regulations, Title 29, Chapter XVII, Part 1910.94d (1975).
8. ACGIH Committee on Industrial Ventilation. *Industrial Ventilation—A Manual of Recommended Practice*, 17th Ed. (Lansing, Michigan: American Conference of Governmental Industrial Hygienists, 1982).
9. American National Standard Z 9.2-1971. "Fundamentals Governing the Design and Operation of Local Exhaust Systems," (New York, New York: American National Standards Institute, 1972).
10. OSHA General Industry Safety and Health Regulations, U.S. Code of Federal Regulations, Title 29, Chapter XVII, Part 1910.1006 (1975).
11. Burchsted, C.A. and A.B. Fuller. "Design, Construction and Testing of High-Efficiency Air Filtration Systems for Nuclear Application," U.S. Atomic Energy Commission, Report No. ORNL-NSIC-65 (Oak Ridge, Tennessee: U.S. AEC Division of Technical Information, 1970).
12. Sax, N.I. *Dangerous Properties of Industrial Materials* (New York, New York: Van Nostrand Reinhold Company, 1975).

13. Graboury, J.A.M. *Conversion Factors for Engineers and Chemists* (Montreal, P.Q., Canada: J. Graboury, 1949).

14. Catalog, Flanders Filters, Inc., Washington, North Carolina (1975).

15. American National Standard N 101.1-1972. "Efficiency Testing of Air-Cleaning Systems Containing Devices for Removal of Particles" (New York, New York: American National Standards Institute, 1972).

16. Ettinger, H.J., J.C. Elder, and M. Gonzales. "Performance of Multiple HEPA Filters Against Plutonium Aerosols," LASL Progress Report No. LA 5349-PR (Los Alamos, New Mexico: University of California—Los Alamos Scientific Laboratory, 1973).

CHAPTER 11

Saving Ventilation Dollars

So far the emphasis has been on designing a ventilation system that works properly. Airflow fundamentals, hood selection, and system design all help to assure a system that, once installed, reduces contaminant levels sufficiently. But how about cost? Today's high equipment and power costs make the money to buy and operate ventilation systems, like other plant equipment, hard to find. Energy price increases make conservation of electrical energy and fuels important industrial goals. This chapter looks at the economics of ventilation systems and ways to protect employee health as well as minimize the costs associated with exhaust ventilation.

VENTILATION COST FACTORS

What are the sources of expenditures in industrial exhaust systems? They are:

- Capital cost for the hoods, ducts, air cleaner, and fan, including installation.
- Power cost for operating the fan and air cleaner. Fan efficiency has a major bearing on operating costs.
- Cost of heating and cooling the air that replaces the air exhausted from the work area.
- Maintenance of the fan; labor to clean out ducts, service the air cleaner, and other maintenance costs.

The size of the ventilation system is important when looking for potential cost savings. A system with relatively low airflow (several thousand ft³/min or less) and moderate static pressure does not cost a lot to operate. But larger systems are expensive to build and operate, especially at higher static pressures typical of systems with filters or other air cleaners.

Techniques for calculating or estimating each of these cost factors will be discussed later in this chapter.

REDUCING AIRFLOW SAVES MONEY

Notice that the first three cost factors mentioned on the preceding page are directly related to the amount of air flowing through the ventilation system. Certainly a larger fan and larger diameter ducts are needed for a system handling a larger airflow. Likewise, the power costs to spin a larger fan will be greater, and since more air is exhausted from the workplace, more make-up or replacement air must be heated or cooled. Thus, the first rule in saving money on a ventilation system is to reduce airflow rates to a reasonable minimum value.

The best way to reduce volumetric flow rate is by proper hood selection. Once the hood design is chosen, the airflow needed to control contaminants is dictated by the hood type, size, and/or distance from the source of contamination. You cannot arbitrarily reduce the airflow into the hood to save money and expect the system to work properly. Instead look closely for a hood design that will control contaminants using the minimum flow while still allowing adequate access and flexibility for the operator. Often you will discover that an enclosing hood with very small open areas or a capturing hood located very close to the process results in the lowest airflow requirements. For example, consider the open surface tank ventilation design example in Chapter 3 (OSHA Ventilation Standards). In this typical case, designing a capturing hood (Figure 11.1) instead of an overhead canopy hood reduces airflow requirements by 75% from 5250 ft³/min to 1350 ft³/min.[1]

The method of minimizing airflow to avoid wasting money depends on whether you are dealing with a new or existing ventilation system.

Reducing Airflow in Existing Systems

The first step in reducing ventilation costs in an existing ventilation system is a thorough survey to locate each hood and determine what type it is (capturing, enclosing, or receiving hood as outlined in Chapter 6) and what the airflow should be. Take the ACGIH *Industrial Ventilation Manual*[2] along and use the hood design diagrams and design criteria. Next measure the airflow at each hood and the capture velocity or face velocity according to testing procedures contained in Chapter 12. This information tells you where you are and is the starting point for efforts to reduce airflow.

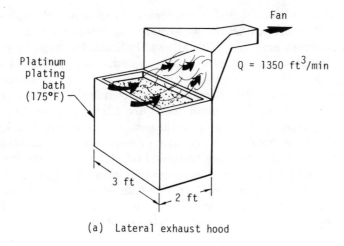

(a)   Lateral exhaust hood

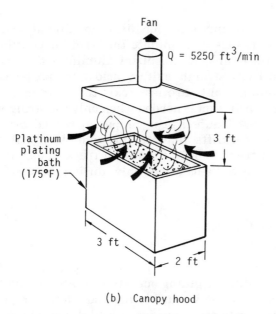

(b)   Canopy hood

Figure 11.1 Comparison of airflow rates to meet Federal OSHA standards for lateral exhaust hood and canopy hood. For the heated platinum plating bath, the lateral hood requires only 25% as much airflow as the canopy hood.

Eliminate Unneeded Ventilation

While you are inspecting the hoods, determine whether each hood is needed to protect health or to satisfy a specific OSHA standard. Remember that, unless a specific requirement exists in OSHA, ventilation is needed only when a hazard assessment indicates that exposures cause a health problem or exceed the allowable limit. Perhaps in your plant some ventilation hoods or even whole systems were installed years ago and changes in plant processes eliminated the need for ventilation. In many laboratories the fume hoods are not used for laboratory procedures requiring ventilation. Unneeded hoods are often used for chemical storage or other purposes that do not require as much airflow as a laboratory fume hood. These hoods can actually present a health hazard because they are often not tested or maintained and may not be working properly when a worker uses one for a rare operation involving toxic materials. It is safer and less expensive to remove these hoods from service, and either dismantle them or post them with out-of-service warning signs. Air sampling may be helpful in identifying unneeded ventilation systems.

Modify Processes

Another potential cost saving is the modification of plant operations or processes in order to reduce or eliminate the need for ventilation. For example, in open surface tanks for anodizing aluminum either a chromic acid-sulfuric acid bath or a straight sulfuric acid bath may be used. Since the sulfuric acid contaminants are less toxic than chromic acid mist, about one-third less ventilation is needed when only the straight sulfuric acid anodizing bath is used. Substituting propane-powered forklift trucks for gasoline-powered forklifts reduces the general ventilation rate from 8000 ft³/min per truck to 5000 ft³/min per truck. If electric forklifts are practical, they need ventilation only during battery charging operations.[3]

Replace Dilution Ventilation With Local Exhaust

Consider replacing dilution ventilation with local exhaust ventilation if practical. For some coating or gluing operations that cover a large area, dilution ventilation is the only feasible method. However, if make-up air heating or cooling costs are excessive, a well-designed local exhaust may do the job at lower cost.

Correct Poor Fan Inlets

You cannot afford to give away 20% of the fan's capacity with a poor fan inlet connection that causes spinning or uneven airflow into the fan. Since spinning or uneven flow reduces the fan's ability to move air, the fan will not perform up to its rated capacity. Other system design features that add to the operating cost are sharp duct elbows, narrow duct diameters, and small openings or slots in hoods that increase friction and turbulent resistance to air movement through the system. For many existing systems it is too late to reduce airflow resistance in the ducts since the system was designed with that amount of pressure loss. Reducing the resistance in these ducts may either have no effect or unbalance the airflow to other hoods. However, it is never too late to correct poor fan inlet connections.

For example, consider the ventilation system segment shown in Figure 11.2. There is an elbow in the fan inlet duct. In Figure 11.2a the elbow is located about 5 duct diameters from the fan inlet. If the elbow was located just before the fan (Figure 11.2b) instead, the fan would lose about 12% of its volume output capability due to turbulence at the fan inlet. To achieve the desired output the fan static pressure would have to be increased about 30 percent.[4] For a moderate size fan (say 10,000 ft^3/min at 9 in. of water fan static pressure), a 35-horsepower motor for the fan would be needed with the poor fan inlet arrangement in Figure 11.2b. A 25-horsepower motor for the fan would be needed with the inlet connection in Figure 11.2a.

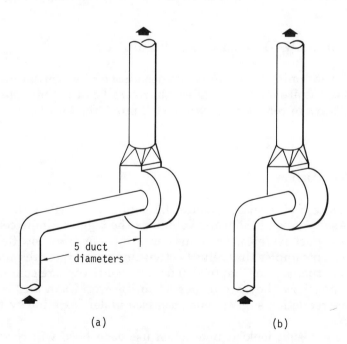

5 duct
diameters

(a)   (b)

Figure 11.2 Locating an elbow several duct diameters away from the fan inlet (a) preserves fan performance. An elbow at the fan inlet (b) can reduce the fan's static pressure capacity by about 30%.

Recirculate Exhausted Air

In every ventilation system the exhausted air must be replaced by make-up air. In order to eliminate drafts and slamming doors as air infiltrates cracks and other openings, this air is best supplied by a fan or other system. Occasionally insufficient make-up air interferes with fan performance, especially with propeller fans used for general ventilation.

Whether air is supplied through a separate make-up system or not, the make-up air is heated or cooled to the workroom temperature. With a supply system the air-tempering can occur before the air is discharged into the room. Without a make-up system the air that infiltrates the workroom is tempered as it mixes with the air already in the room.

For large ventilation systems the cost of heating and cooling the make-up can be prohibitive. Recirculation of cleaned exhaust air is attractive for these large systems (Figure 8.14). The advantages and disadvantages of recirculating exhausted air are covered in Chapter 8.

Another way to reduce make-up air heating or cooling costs is to supply untempered make-up air into booths or enclosures. For example, a supplied-air laboratory hood (Figure 11.3) using outdoor air directed into the hood can reduce the amount of room air exhausted by 70%.[5]

DESIGNING NEW SYSTEMS TO SAVE MONEY

During the preliminary planning and design of a new ventilation system, many decisions are made that affect the overall cost of the system. Here are some ways to reduce costs while still providing for employee safety and health.

Reduce Airflow

The hood size and location dictate the volumetric airflow through the system. As discussed earlier air volume is the biggest single cost factor in many exhaust systems. Try to use enclosures or other low flow hoods if possible. Air samples that reflect actual employee exposures and smoke tube tracer studies (see Chapter 12) help to identify where contaminants originate and how they are dispersed in the workroom air.[3] Review the airflow reduction suggestions discussed under "existing systems" in this chapter.

During design look at how much use each hood will receive. For example, it is wasteful to continuously exhaust large amounts of air through a large soldering hood that is hardly used but is on the same fan system as a grinder used often throughout the day. During design it may be more economical to install a smaller fan on each unit rather

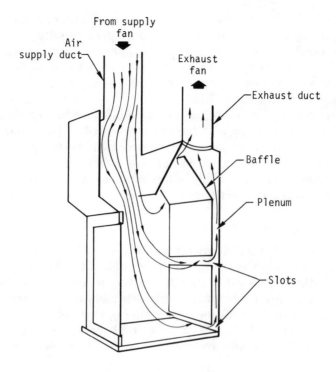

Figure 11.3 A laboratory hood with a supply system
reduces the make-up air needed for the laboratory
when the sash is lowered.

than one large fan for both hoods. Occasionally a process permits design
of a ventilation system that exhausts only part of the hoods at any time.
Dampers are closed when the machinery or process is not operating and
they are opened again when needed. The potential disadvantages of these
partial systems are obvious: if airflow to the unused hoods is not turned
off then airflow to all hoods will probably be insufficient. Also workers
must remember to open the dampers when needed. Interlocked controls
between the process and damper eliminate these problems but add to
capital and maintenance expense.

Reduce Airflow Resistance

With all other factors constant, reducing the resistance or pressure loss
in the system decreases the operating cost. If the system has an air
cleaner, it may be the single largest source of resistance. Although some
air cleaning devices cause low or moderate resistance, high-energy scrub-

bers, filters, or carbon beds can cause 5–10 in. of water or more pressure loss. Oversizing the air cleaner may reduce pressure drop.

High hood slot or duct velocities also cause high pressure losses. Friction and turbulent pressure losses vary with the square of velocity. If you double the velocity through a hood opening or duct, the pressure loss increases by a factor of four.

The problem with applying the principle of reducing airflow resistance is that many other cost factors are linked to duct diameter and air velocity. This means that reducing operating costs by lowering system resistance may increase another cost factor. Although the system must be considered as a whole,[6] there are general ways to reduce static pressure that can be investigated.

Long Runs of Narrow Ducts

Avoid long runs of small-diameter ducts or other ducts with high pressure loss.[7] The high pressure drop through these ducts often makes them the path of greatest resistance in the entire system. Remember from Chapter 8 that the branch of greatest resistance governs the final fan static pressure. Few of the money-saving tips discussed earlier will be successful once a system with long, high pressure loss ducts is installed. Instead divide the system into two or more separate systems, each with a fan.

Example: A ventilation system will be installed to exhaust three soldering hoods and a pedestal grinder as shown in Figure 11.4. Is there any advantage in dividing the system into two separate systems?

Answer: By applying the design methods described in Chapter 8 the fan for the large single system will draw 2000 ft³/min at 8.8 in. of water fan static pressure. According to the fan rating table in Figure 11.5, the fan will need a 4.0-horsepower motor. By separating the two systems as shown in Figure 11.6 the fan sizes will be 1500 ft³/min at 4.24 in. of water fan static pressure for the three soldering hoods and 500 ft³/min at 6.1 in. of water for the grinder. The motor sizes will be 1.5 and 1.25 horsepower respectively for a total of 2.75 horsepower for the separate systems.

Locate Fan Near High Suction Ducts

If ducts with high static pressure are unavoidable, locate the fan so that these branch ducts enter the main duct as close to the fan as possible.[7] The reason for this is that the suction or negative static pressure is greatest near the fan inlet. High pressure loss ducts need greater suction to overcome resistance; by locating these ducts near the fan, the suction throughout the system does not have to be increased so that these ducts will work properly. In most systems the reason for using a narrow duct is to achieve the transport velocity to prevent dust settling in the ducts; in gas and vapor systems high pressure loss ducts should be avoided.

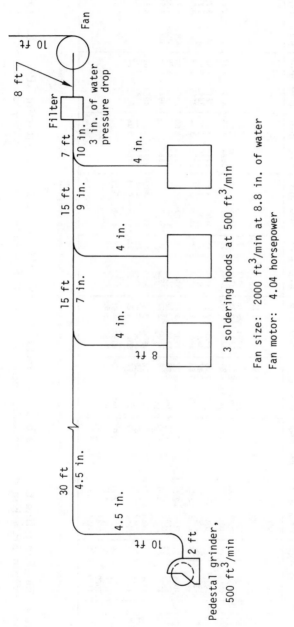

Figure 11.4 A typical local exhaust system showing the duct diameters and fan size. This system is used to illustrate the effect of modifications on fan size and power costs.

CFM	4" SP		4½" SP		5" SP		5½" SP		6" SP		6½" SP		7" SP		7½" SP		8" SP		8½" SP	
	RPM	BHP	RPM	BHP	RPM	BHP	RPM	BHP	RPM	BHP	RPM	BHP	RPM	BHP	RPM	BHP	RPM	BHP	RPM	BHP
1289	2699	1.24	2841	1.42																
1375	2730	1.32	2868	1.49	3001	1.68														
1461	2762	1.39	2895	1.57	3024	1.76	3160	1.95	3277	2.16										
1547	2797	1.47	2927	1.66	3052	1.85	3180	2.05	3299	2.26	3417	2.47								
1633	2836	1.55	2963	1.75	3086	1.94	3206	2.15	3323	2.35	3438	2.57	3551	2.79	3660	3.01				
1719	2878	1.64	2999	1.84	3119	2.04	3238	2.25	3356	2.47	3465	2.68	3576	2.91	3684	3.14	3788	3.37	3892	3.61
1805	2922	1.73	3042	1.94	3160	2.15	3273	2.36	3386	2.58	3498	2.81	3603	3.03	3710	3.26	3812	3.51	3916	3.75
1891	2969	1.83	3087	2.04	3201	2.25	3312	2.47	3424	2.71	3530	2.93	3635	3.16	3740	3.41	3843	3.64	3940	3.89
1977	3023	1.94	3137	2.15	3245	2.37	3355	2.59	3462	2.82	3567	3.06	3671	3.31	3771	3.54	3868	3.78	3970	4.04
2063	3075	2.04	3185	2.26	3290	2.48	3398	2.72	3503	2.95	3604	3.19	3707	3.44	3807	3.69	3903	3.94	4000	4.21
2149	3131	2.16	3238	2.39	3340	2.61	3441	2.84	3543	3.08	3644	3.33	3746	3.58	3842	3.83	3938	4.09	4030	4.35
2235	3187	2.28	3290	2.51	3394	2.75	3493	2.98	3590	3.22	3693	3.49	3788	3.74	3881	3.99	3973	4.25	4065	4.51
2407	3306	2.54	3404	2.78	3501	3.02	3599	3.28	3690	3.53	3786	3.81	3879	4.07	3964	4.31	4059	4.61	4143	4.87
2579	3432	2.82	3524	3.07	3617	3.33	3708	3.59	3799	3.86	3888	4.13	3975	4.41	4059	4.68	4147	4.97	4232	5.26
2751	3559	3.13	3651	3.41	3738	3.67	3825	3.94	3910	4.21	3998	4.51	4079	4.78	4164	5.08	4245	5.37		
2923	3692	3.46	3779	3.74	3866	4.03	3947	4.31	4032	4.61	4110	4.89	4191	5.19						
3095	3830	3.83	3912	4.12	3994	4.42	4075	4.72	4154	5.02	4232	5.32								
3267	3969	4.23	4047	4.53	4130	4.85	4206	5.16												
3439	4110	4.66	4191	4.99	4267	5.31														

Selection within tinted area renders most efficient, quietest operation.   BHP shown does not include belt drive losses.

Figure 11.5 Fan rating table for a backward inclined blade fan. CFM = Flow rate, ft³/min; RPM = Rotating speed, rev/min; SP = Fan static pressure, inches of water; BHP = Brake horsepower. (Source: Reference 10)

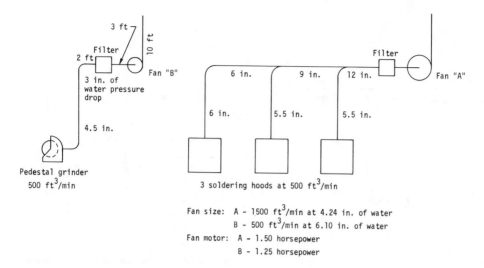

Fan size: A - 1500 ft³/min at 4.24 in. of water
           B - 500 ft³/min at 6.10 in. of water
Fan motor: A - 1.50 horsepower
           B - 1.25 horsepower

Figure 11.6 The exhaust system in Figure 11.4 divided into two separate systems to reduce power costs.

Example: For the system shown in Figure 11.4 evaluate the effect of locating the fan near the pedestal grinder rather than near the soldering hoods.

Answer: The proposed system is shown in Figure 11.7. By designing both systems according to the methods in Chapter 8, the system with the fan located by the soldering hoods (Figure 11.4) is 2000 ft³/min at 8.8 in. of water fan static pressure. If the system is designed with the fan near the grinder, the fan size will be 2000 ft³/min at only 6.0 in. of water fan static pressure.

Building the system as shown in Figure 11.7 will reduce operating expenses, since the electrical consumption will be about 75% of that of the system in Figure 11.4. However, the construction cost will be higher since 30 ft of 8-in.-diameter duct rather than 4.5-in. duct is needed to connect the soldering hoods with the grinder hood if the fan is located near the grinder.

As shown by this example, the different cost factors must be balanced to give the most economical overall system.

Balance Capital and Operating Costs

Designing a new system makes it easier to examine the relationship between operating and capital costs. With an existing system the ducts and other hardware are already installed. Justifying replacement is difficult unless a large operating cost savings is possible. For a new system look at the whole system when balancing capital and operating costs. Keep in mind that fan costs (capital and operating) increase with high

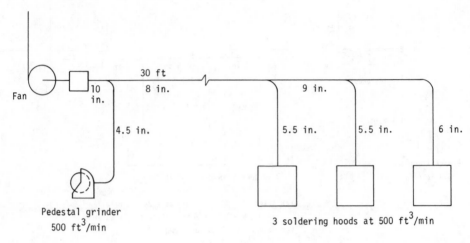

Figure 11.7   The exhaust system in Figure 11.4 redesigned to save power costs
even though the capital cost for the duct is increased.

airflow and resistance, or static pressure; duct capital costs rise as you
increase duct diameters to lower resistance in the system. There is a
range of optimum duct sizes for every ventilation system where overall
cost is lowest. Economics is not the sole factor in determining the best
duct size. For dust ventilation systems a minimum duct velocity, usually
3500–4500 ft/min, must be maintained to keep dust from settling. As a
general rule duct velocities of 1800–2500 ft/min give a good balance
between duct installation cost and fan operating costs when this velocity
does not cause settling problems in the ducts.[8] The increase in resistance or fan static pressure with different duct sizes can be estimated if
the static pressure for any specific duct diameter has been calculated.
Since friction and turbulent losses increase approximately as the square
of velocity, apply this equation for estimating the effect of different duct
diameters:

$$Q = V \times A$$

The duct cross-sectional area can be calculated from:

$$A = \frac{\pi D^2}{4} \tag{11.1}$$

where Q = volumetric airflow, ft³/min
V = duct velocity, ft/min
A = duct cross-section area, ft²
D = duct diameter, ft

To consider different duct diameters in the same system, Equation 11.1 becomes:

$$Q = \frac{V_1 \pi D_1^2}{4} = \frac{V_2 \pi D_2^2}{4} \tag{11.2}$$

$$\frac{V_2}{V_1} = \frac{D_1^2}{D_2^2} \tag{11.3}$$

Since pressure loss varies with the square of velocity,

$$SP_2 = SP_1 \left(\frac{V_2}{V_1}\right)^2 = SP_1 \left(\frac{D_1^2}{D_2^2}\right)^2 \tag{11.4}$$

$$SP_2 = SP_1 \left(\frac{D_1}{D_2}\right)^4 \tag{11.5}$$

where $SP_{1,2}$ = static pressure with different duct diameters, in. of water

So changes in resistance to airflow through the system vary inversely as the fourth power of changes in duct diameter. This equation is for estimating only; static pressure should be calculated using techniques in Chapter 8 for accurate results.

Example: In Chapter 8 a welding bench system was designed to draw 1050 ft³/min through an 8.5-in.-diameter duct (Figure 11.8). The calculated fan size was 1050 ft³/min at 0.86 in. of water fan static pressure. If the duct diameter was reduced to 5 in. to save on duct material costs, what is the new fan size for the system?

Answer: The flow will remain constant (1050 ft³/min) but a new fan static pressure has to be calculated to overcome the added resistance. Using Equation 11.5:

$$SP_2 = SP_1 \left(\frac{D_1}{D_2}\right)^4 = 0.86 \text{ in. of water} \left(\frac{8.5 \text{ in.}}{5.0 \text{ in.}}\right)^4$$
$$SP_2 = .86 \ (8.35) = 7.2 \text{ in. of water}$$

So you will need a fan rated at 1050 ft³/min and 7.2 in. of water fan static pressure. This fan will require a 2.5-horsepower motor rather than the quarter horsepower motor that would be satisfactory if the 8.5-in. ducts were installed.

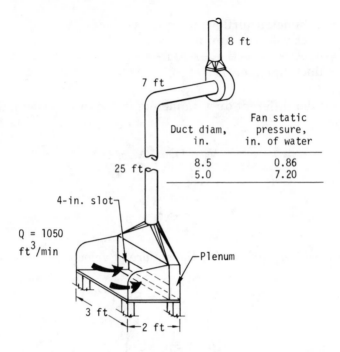

Duct diam, in.	Fan static pressure, in. of water
8.5	0.86
5.0	7.20

Figure 11.8   Reducing the duct diameter from 8.5 inches to 5 inches increases the required fan static pressure by almost 900%.

SYSTEMS WITH VARYING AIRFLOW

Some ventilation systems operate at reduced airflows for considerable periods of time. If the system has an air cleaner that gradually plugs, the system is usually designed to draw more air than required when the device is clean so that minimum flow criteria are satisfied as the air cleaner plugs before cleaning or replacement. In other ventilation systems parts of the process equipment are used infrequently or are taken out of service for long periods for maintenance or repair. One traditional way to adjust the airflow in systems with varying air requirements is to install dampers in the ducts to provide an artificial source of pressure loss that reduces airflow when the air cleaner is not plugged or when some hoods are disconnected. Another technique is to exhaust more air than needed at some times to meet minimum flow criteria during periods of higher demand or resistance.

Operating costs can be reduced by installing either fan inlet dampers or a motor speed controller in the system to reduce the amount of work the fan does on the air, and so its power consumption, during periods of low flow.[9]

Fan Inlet Dampers

Inlet dampers, or vanes, are triangular pieces of metal installed near the fan inlet (Figure 11.9) that impart a spin to air entering the fan. The spin is in the direction the fan is rotating and reduces the amount of work the fan does on the air since the fan blades have to "catch up" to the moving air. This means that the fan will not move the maximum amount of air, and power consumption will be reduced. The inlet dampers modify the static pressure curve for the fan to move it below the operating curve with no inlet dampers. Inlet dampers are available for either manual or automatic adjustment as the system resistance changes. An example later in this section illustrates their use.

Fan Speed Controls

The way to minimize operating costs in a system with varying airflow is to install a speed control on the fan motor. Changing the fan rotating speed moves the fan operating point to a completely new fan curve as

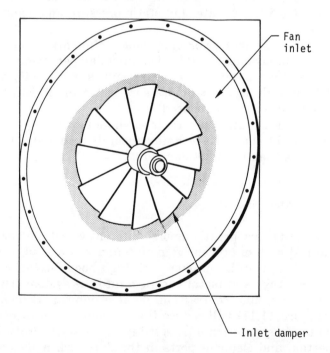

Figure 11.9 Fan inlet dampers, or vanes, impart a spin to air entering the fan and reduce both the fan's output and power consumption.

explained in Chapter 9. This provides operating cost savings that exceed the savings from fan inlet dampers. Although the cost of motor speed controls was once high, moderate cost solid state controllers for both alternating and direct current motors are now available.

Fan speed can also be controlled on belt-driven fans by installing different diameter pulleys on the fan and motor, and changing the fan belt to the correct size pulleys to produce the desired speed. This is a high labor cost operation if done frequently.

The effect of fan inlet dampers and fan speed controls on fan rating curves is illustrated in Figure 11.10. Assume that these curves apply to a system that contains a filter and requires 6000 ft³/min to operate properly. The fan static pressure varies from 2.5–5.5 in. of water depending on whether the filter is clean or dirty. The goal is to maintain the 6000 ft³/min flow rather than 9000 ft³/min when the filter is clean. There are three main choices:

- Install an adjustable blast gate or duct damper to create an artificial source of resistance or pressure drop (Figure 11.10a). With a clean filter, an additional 3.0 in. of water pressure drop from the damper is needed. This works in reducing the airflow but the fan is still operating at 5.5 in. of water fan static pressure and about 6 horsepower.
- Install a fan inlet damper or inlet vanes to impart a spin to the air entering the fan (Figure 11.10b). The inlet damper changes the slope of the fan static pressure curve. The result is that the fan will draw 6000 ft³/min with a fan static pressure of 2.5 in. of water. The fan brake horsepower is also reduced.
- Install fan speed controls to create a whole new fan static pressure curve (Figure 11.10c). This technique usually results in a larger electrical savings than fan inlet dampers.

SAVING ON MAINTENANCE

Maintenance costs vary greatly among different ventilation systems. In any mechanical system preventive maintenance is a good way to save money in the long run. In systems with high hood face velocities it is possible to suck rags and metal turnings into the system. These often lodge in an elbow, duct contraction, or at a damper. Screening hood openings (Figure 11.11) will reduce these problems. Duct systems are susceptible to plugging from dust settling in the duct. Providing sufficient inspection and cleanout ports in the ductwork makes it easy to locate and remove blockages.

Many ventilation hoods are added on to existing machinery. If the hoods protrude too far out from the equipment, they may be prone to

Figure 11.10 There are three ways to change the airflow in systems with varying pressure loss or airflow requirements: (a) duct dampers add artificial pressure loss but do not reduce fan power consumption; (b) fan inlet dampers change the shape of the static pressure curve and save power; and (c) fan motor speed controls create a new static pressure curve and usually result in the greatest power saving. (Source: Reference 9)

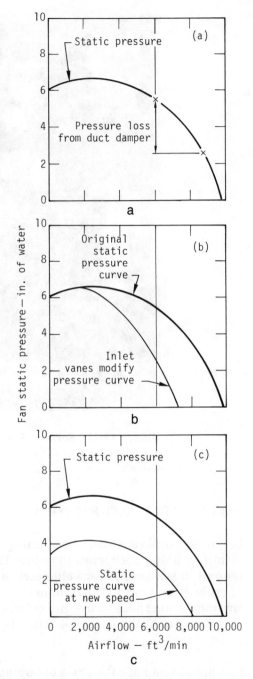

damage from overhead cranes and forklifts used in the plant. Minimize these repair costs by careful placement and shielding of ventilation hoods and ducts.

Figure 11.11   Screened hood openings reduce maintenance costs that are due to debris-clogged ducts.

CALCULATING COST FACTORS

So far the different costs of running a ventilation system have been identified along with ways to reduce these costs. Cost savings were expressed in terms of lower volumetric airflow, lower fan static pressure, or brake horsepower. The remainder of this chapter contains guidelines for translating these terms into dollars and cents for an economic study.

The aim of the economic study is to balance capital and operating costs:

- Capital costs are the price of buying and installing the hoods, ducts, air cleaner, fan, and the make-up air system. These costs can be minimized by using good judgment in simplifying duct layout, using short duct lengths, reducing airflow volume, and choosing an air cleaner with low resistance.

- Operating expenses are the cost of electricity to spin the fan, supplying heated or cooled make-up air (or tempering the air that infiltrates the workroom if no make-up system is provided), operating the air cleaner if it requires power, and the maintenance on the entire system including the air cleaner.

The importance of airflow volume and system resistance on capital and operating costs was highlighted earlier. Table 11.1 summarizes the impact of airflow and resistance on the different cost factors. All cost factors except hood costs decrease or are unchanged by lowering the air volume. It costs more to provide hoods that are shaped and located to minimize the airflow needed for proper operation than it costs to provide less effective hoods. Likewise efforts to reduce system resistance lower or do not change all of the cost factors except duct costs and perhaps air cleaner costs. To reduce system resistance, duct diameters are usually increased, wider duct elbows are installed, turning vanes are added to elbows, and angled rather than perpendicular branch duct entries into main ducts are specified. Also gradual rather than abrupt duct enlargements and contractions are needed. Air cleaner costs may increase because, if the air cleaner is a major source of resistance, an oversized or a different type air cleaner may be needed to reduce resistance. Either of these air cleaner choices will probably cost more than the minimum air cleaner.

Duct Costs

Duct costs depend on the amount and kind of material used for duct construction and the labor to fabricate and install the ducts. Steel sheet

Table 11.1 System Cost Factors Compared with Airflow and Resistance

	Effect of Reducing Airflow Needed for Proper Operation	*Effect of Reducing System Resistance*
Capital Costs		
Hoods	Increases	No change
Ducts	Decreases	Increases
Air cleaner	Decreases	May increase
Fan	Decreases	Decreases
Make-up air system	Decreases	No change
Operating Costs		
Power for fan	Decreases	Decreases
Maintenance	No change	No change
Heating/cooling Make-up air	Decreases	No change

is used in most duct systems although stainless steel, Polyvinyl chloride, and other materials are popular in corrosive or special environments. The amount of steel used in ducts increases sharply with large-diameter ducts (Figure 11.12). A 1-ft-diameter duct weighs about 500 lbs per 100 ft of duct while a 3-ft-diameter duct weighs about 2500 lbs per 100 ft. That is a lot of steel to buy, fabricate, and hang on the rafters. Figure 11.12 is developed from recommendations in the ACGIH *Industrial Ventilation Manual*[2] for systems carrying moderately abrasive materials or light concentrations of highly abrasive materials: ducts up to 8 in. are 22-gauge steel; from 8–18 in. are 20-gauge; 18–30 in. are 18-gauge and over 30 in. are 16-gauge steel. Ducts carrying other contaminants may require other duct thicknesses.

Labor costs vary between location and with time and so should be investigated locally for cost estimates.

There is little operating cost associated with ducts except the need for cleaning when plugging occurs.

Fan Costs

Fan capital costs vary with air volume and fan static pressure. For moderately sized systems (<10,000 ft³/min and 10 in. of water fan static

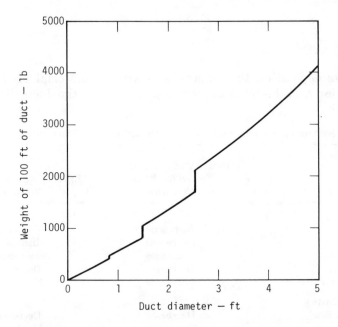

Figure 11.12 Duct weight as a function of duct diameter. Larger ducts require thicker sheet metal, which accounts for the steep increases in weight.

pressure), the potential capital cost savings from designing a system with slightly lower volume or fan static pressure is not significant. The cost savings are realized from long-term lower electrical power consumption in operating the fan.[10]

Operating costs are calculated from the fan brake horsepower. The formula and sample calculation are in the "Power Cost" section in this chapter. Figure 11.13 shows the relationship between airflow volume, fan static pressure, and brake horsepower for a backward inclined blade fan, one of the more efficient types of centrifugal fans. As shown in Figure 11.13 the fan static pressure has a profound influence on power consumption. The slope of the volume versus horsepower curve increases sharply at higher fan static pressures. Of course a higher horsepower requirement means power costs to operate the fan will be higher.

Fan efficiency is defined as the amount of air volume the fan will move against resistance for a given electrical power consumption. Although a fan rated to deliver 2000 ft³/min at 5.0 in. of water fan static pressure will meet these criteria, a more efficient fan will do it with less electrical power consumption. Figure 11.14 shows the same information as in Figure 11.13 except for a radial blade fan.[10] The radial blade is less efficient than other types of centrifugal fans but has good resistance to abrasion and clogging and so is used in systems carrying high con-

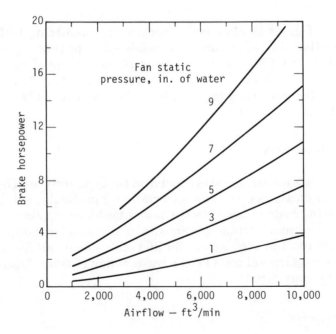

Figure 11.13 Brake horsepower as a function of airflow and fan static pressure for a backward inclined blade fan.

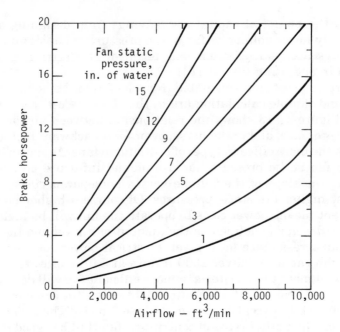

Figure 11.14   Brake horsepower as a function of air-flow and fan static pressure for a radial blade fan.

centrations of dusts or abrasive materials. By comparing both figures you can see that the radial blade fan is less efficient than the backward inclined blade fan. For example, to move 10,000 ft³/min of air at 7 in. of water fan static pressure, the radial blade fan needs a 21-horsepower motor while the backward inclined blade fan requires only a 15-horsepower motor.

Air Cleaner Costs

The capital costs for air cleaners depend on the type and size. The higher-efficiency air cleaners cost more, especially for particulate air cleaners.

Operating costs include electric power for blowers, pumps and electrostatic precipitators; replaceable media in filters and carbon adsorption beds; fuel for incinerators; and water and other chemicals for scrubbers. Labor and final disposal costs for the waste removed from the air cleaner must also be considered.

Power Costs

Electricity is the usual power source to spin the exhaust fan, the make-up air fan, and the air cleaner when it requires power. For fans the power costs can be calculated from the brake horsepower rating of the fan motor. By definition:

$$1 \text{ horsepower} = 745.7 \text{ watts} = 0.75 \text{ kilowatt}$$

Operating a 5-horsepower motor for 1 hour consumes 3.75 kilowatt-hours of electricity. Brake horsepower is the actual energy needed to spin the fan including friction losses, although the losses in the drive (belt or direct drive) are not included in the brake horsepower values in fan rating tables.

Power costs vary with the utility company supplying the power and the class of service. Although fuel shortages make future power costs projections shakey, they will undoubtably increase over the forseeable future.

Example: A radial blade fan rated at 6000 ft³/min and 7.0 in. of water fan static pressure requires a 12-horsepower motor. The fan runs 16 hr/day for 5 days per week. If the electrical rate is 8 cents per kw-hour, what is the yearly power cost?

Answer: In a year the fan operates:

$$\frac{16 \text{ hr}}{\text{day}} \times \frac{5 \text{ days}}{\text{week}} \times \frac{52 \text{ weeks}}{\text{yr}} = 4160 \text{ hr/yr.}$$

The fan uses:

$$12 \text{ HP} \times \frac{0.75 \text{ kilowatt}}{\text{HP}} = 9.0 \text{ kw/hr.}$$

So the yearly cost is:

$$\frac{9.0 \text{ kw}}{\text{hr}} \times \frac{4160 \text{ hr}}{\text{yr}} \times \frac{\$0.08}{\text{kw}} = \$2995.20/\text{yr.}$$

Example: A backward inclined fan will move 6000 ft³/min at 7.0 in. of water fan static pressure with a 9-horsepower motor. Assuming the same operating and power cost as above, what is the yearly cost:

Answer: The fan uses:

$$9.0 \text{ HP} \times \frac{0.75 \text{ kilowatt}}{\text{HP}} = 6.75 \text{ kw/hr.}$$

So the yearly cost is:

$$\frac{6.75 \text{ kw}}{\text{hr}} \times \frac{4160 \text{ hr}}{\text{yr}} \times \frac{\$0.08}{\text{kw-hr}} = \$2246.40/\text{yr.}$$

If a backward inclined fan could be used rather than a radial fan, $748.80 per year in power costs could be saved.

Heating and Cooling Make-Up Air

For the ventilation system to work properly, sufficient make-up or replacement air must be provided. Generally 90–110% of the exhaust air volume is supplied if the plant has a forced make-up air system. With no planned system for replacement air, make-up air enters the building through open doors or windows and through cracks around doors and windows. Either way, the incoming air will be heated or cooled by the plant heating or air conditioning system. The advantage of supplying tempered make-up air to the work area is that workers may not use ventilation systems when needed during cold weather to avoid the chilling drafts from infiltrating outdoor air. Some state OSHA regulations require that make-up air have a temperature of no less than 65°F although the federal OSHA regulations contain no standards.

Heating Costs

The heat required is calculated from this equation:[3]

$$H = Q \times 0.075 \times 0.24 \times 60 \, (t_d - t_o) \qquad (11.6)$$

where H = heat required, Btu/hr
  Q = volumetric airflow, ft³/min
  0.075 = density of standard air, lb/ft³
  0.24 = specific heat of air, Btu/lb/°F
  60 = converts Btu/min to Btu/hr
  t_d = room air temperature, °F
  t_o = outside air temperature, °F

The heat required (H) is the actual heat to be added to the air and must be adjusted for the efficiency of the heating device. The efficiency of electrical heaters and gas and oiled fired heaters is 65–80%.

The yearly heating cost can be estimated from the number of degree-days per year at the plant location. The degree-day rating for a given day is the number of degrees that the average daily temperature was below the reference temperature, usually 65°F. The annual degree-days is the summation of degree-day ratings for different cities in the United States (Appendix D). The yearly heating cost is calculated from this equation.[2]

$$C = \frac{0.154 \times Q \times T \times DD \times CF}{q} \qquad (11.7)$$

where C = annual heating cost, dollars
  Q = volumetric airflow, ft³/min
  T = operating time, hr/week
  DD = annual degree-days (Appendix C)
  CF = cost of fuel, $/unit
  q = available heat per unit of fuel (Table 11.2)

Table 11.2   Approximate Heat Available from Fuels

Fuel	Available Heat/Unit
Coal	6000 Btu/lb
Oil	100,000 Btu/gal
Gas	800 Btu/ft³

Source: Reference 2.

Example: What is the approximate yearly cost of heating 9000 ft³/min to 65°F for 40 hr/ week using oil that costs $1.15/gallon in Pittsburgh, Pennsylvania?

Answer: From Appendix C a normal year has 5930 heating degree-days assuming a discharge temperature of 65°F. From Table 11.2 there is about 100,000 Btu of heat available per gallon burned.

Applying Equation 11.7:

$$C = \frac{0.154 \times Q \times T \times DD \times CF}{q}$$

$$= \frac{0.154 \times 9000\, \frac{ft^3}{min} \times \frac{40\ hrs}{wk} \times 5930\, \frac{DD}{yr} \times \frac{\$1.15}{gal}}{100{,}000\ Btu/gal}$$

$$C = \$3780.73 \text{ per year}$$

In industrial settings involving heated processes, introducing make-up air that is warmed by waste heat before reaching the worker will reduce the amount of fuel needed.

Cooling Costs

For many applications air conditioning seems to be an extravagance hardly worth considering. However, many work areas are air conditioned for human comfort and work requirements. For example, laboratories may require a controlled environment with temperature and humidity maintained within narrow ranges. Electronic assembly and other clean industries are usually air conditioned.

Cooling make-up air is more expensive than heating it on a degree-for-degree basis. One rule of thumb is that for every 1000 ft³/min exhausted through hoods, 3–4 tons of refrigeration must be added to the air conditioning system.[11] Cost for air conditioning equipment varies; this information should be obtained from local equipment suppliers for use in the cost calculations when the system airflow has been determined. If you are dealing with air conditioned make-up air, every step to reduce the exhaust volume should be considered.

Operating costs are calculated from the size of the air conditioning unit and the electrical power to run it. Design of air conditioning systems involves calculating the sensible heat gain corresponding to heat sources in the area and temperature of incoming air plus the latent heat from the change in humidity between incoming and room air. Calculation methods are contained in the ASHRAE *Handbook of Fundamentals* [12] and other texts.

Electrical power costs can be estimated from Figure 11.15 if you know the cooling capacity required. One ton of air conditioning equals 12,000 Btu/hr of heat removed. The ton is a unit used in air conditioning work because the latent heat (heat removed by melting) of one ton of ice equals about 288,000 Btu and so a ton-day is that amount of heat removed in a day.

Example: Assuming 3 tons of air conditioning are required per 1000 ft³/min of exhausted air, what is the power consumption to cool 3000 ft³/min of air?

Answer: 3000 ft³/min is equivalent to 9 tons.

$$9 \text{ tons} \times 12,000 \frac{\text{Btu/hr}}{\text{ton}} = 108,000 \text{ Btu/hr}$$

From Figure 11.15 108,000 Btu/hr requires about 8 kw-hr of electric power.

Note that the number of hours per year that the system will operate based on temperature, humidity, and other heat sources, and periods of occupancy must be calculated from the ASHRAE *Handbook* before yearly

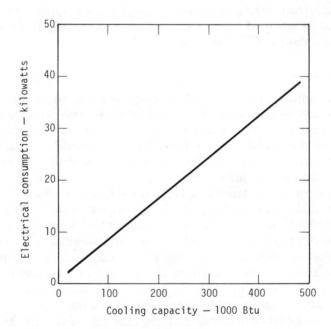

Figure 11.15 Approximate air conditioner electrical consumption as a function of cooling capacity.

power costs can be calculated. Appendix C contains cooling degree-day information for cities in the United States.

SUMMARY

This chapter covered the costs involved in building and operating a ventilation system and ways to minimize these costs. For small systems that draw a few thousand cubic feet per minute or less against low static pressure, the potential for saving money is limited. But for larger systems or systems with high static pressure loss the savings can be significant.

REFERENCES

1. "OSHA General Industry Safety and Health Regulations," U.S. Code of Federal Regulations, Title 29, Chapter XVII, Part 1910.94 (1975).
2. ACGIH Committee on Industrial Ventilation. *Industrial Ventilation—A Manual of Recommended Practice,* 17th Ed. (Lansing, Michigan: American Conference of Governmental Industrial Hygienists, 1982).
3. Hama, G.M. "Conservation of Air and Energy," *Michigan's Occupational Health* 19, No. 3 (1974).
4. Trickler, C.J. "Effect of System Design on the Fan," Engineering Letter No. E-4 (Chicago, Illinois: The N.Y. Blower Company).
5. Catalog. Kewaunee Technical Furniture Company, Statesville, North Carolina.
6. Rogers, A.N. "Evaluation in Fan Selection," Reprint No. 5037 (Pittsburgh, Pennsylvania: Westinghouse Electric Corporation, 1954).
7. National Institute for Occupational Safety and Health. *The Industrial Environment—Its Evaluation and Control* (Washington, D.C.: U.S. Government Printing Office, 1973).
8. Hemeon, W.C.L. *Plant and Process Ventilation* (New York, New York: Industrial Press, Inc., 1963).
9. Burchsted, C.A. and A.B. Fuller. "Design, Construction, and Testing of High Efficiency Air Filtration Systems for Nuclear Application," U.S. Atomic Energy Commission, Report No. ORNL–NSIC–65 (Oak Ridge, Tennessee: U.S. AEC Division of Technical Information, 1970).
10. Catalog, New York Blower Company, Chicago, Illinois.
11. Constance, J.D. "Clearing the Air in Laboratories," *Research/Development* 23, No. 9, 22 (1972).
12. American Society of Heating, Refrigerating and Air-Conditioning Engineers, Inc. *Handbook of Fundamentals* (New York, New York: ASHRAE, 1972).

CHAPTER 12

Testing

Every ventilation system for contaminant control should be checked periodically for proper operation. A newly installed system should be thoroughly tested to see that it meets design specifications and to assure that airflow through each hood is correct, that is, that the system is properly balanced. Periodic testing may also be required by OSHA regulations. Other reasons for testing are listed in Table 12.1.

In order to show the operating conditions (damper setting, fan speed) and location of measurements (Figure 12.1), results of initial tests should be recorded on data forms or a drawing of the system. Airflows or velocity should be compared to OSHA standards or hood design drawings in the ACGIH *Manual*.[1] Results of air sampling for contaminants should be available to evaluate the effectiveness of the overall hood selection, airflow, location, and design.

From the safety and health department's viewpoint, the periodic checks of system performance are important in protecting workers' health. After a system is installed and operating properly, the system's efficiency in containing or capturing contaminants may decrease for many reasons: loose fan belt, unauthorized damper adjustment, dust settled in ducts,

Table 12.1 Reasons for Testing Ventilation Systems

Type of Test	Reason
Complete fan and system test	Document the initial performance of a new system for comparison with design criteria and for future reference.
	Diagnose malfunctions.
	Determine whether system can be expanded or airflows can be increased.
Periodic velocity or hood static pressure tests	Satisfy OSHA standards.
	Determine whether system performance is adequate.
Air sampling for contaminants	Measure employee exposures.
	Determine systems efficiency in controlling contaminants.
	Locate "fugitive" or uncontrolled sources.

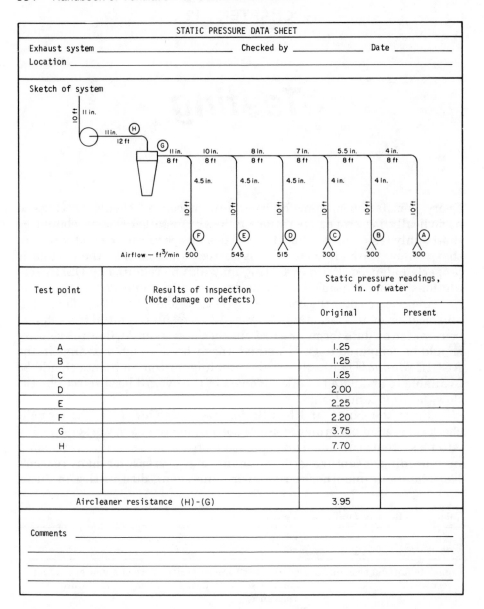

Figure 12.1 Data sheet for recording the results of periodic static pressure tests.

and clogged air cleaners. Occasionally an existing system will become unbalanced when new ducts are added to it. The periodic inspection also gives you the opportunity to observe hood conditions for wear or damage and to see how workers are using the ventilation system.

This chapter describes a variety of different tests that help determine how the system is operating. The tests are arranged in order from simplest to most complex; once you have determined the information you need, you can proceed down the list and select appropriate tests. The tests are: smoke tube tracer studies; velocity at the hood opening; hood static pressure; Pitot tube duct velocity measurements; system static pressure tests; fan pressure, rotating speed, and power consumption; and periodic maintenance.

The important factor is not just how well the hardware is performing but how much protection the system is providing. Although a system's airflow through hoods, air cleaner efficiency, fan speed, and power consumption may check out according to design specifications, the system is not performing adequately if air samples show excessive employee exposures. Perhaps the wrong type or size of hood was specified in the original design or there are uncontrolled sources of contamination in the workplace. Only by linking ventilation system test results with air contaminant levels can you be sure that the system is providing enough protection.

HOOD TESTING TECHNIQUES

These tests are designed to evaluate the airflow into hoods. Both visual tracer tests and quantitative air velocity or pressure tests help determine whether a hood is working properly.

Smoke Tube Tracer Tests

Smoke tubes for testing ventilation systems are available from several manufacturers. These tubes (Figure 12.2) contain titanium tetrachloride or other chemicals that produce fumes by reaction with air blown through the tube. For use, the tube tips are broken off and air is blown through the tube with a rubber squeeze bulb. The smoke follows the air currents and its velocity shows how air is flowing into the hood or in the room. Tracer studies are useful in the following applications:[2]

- Showing the dispersion of contaminants from a source (Figure 12.3a). The correct hood location and size depend on where the contaminants originate and how much velocity they have. Smoke tubes help identify the point where the initial velocity of contaminants away from the hood is dissipated so the airflow to develop the necessary hood capture or containment velocity at that point can be calculated.
- Determining the approximate capture distance for hoods (Figure 12.3b). Smoke tubes show how far from a hood opening the contaminant will be captured. Although smoke tubes are not as accurate as

Figure 12.2   Smoke tubes are used to determine air currents and relative velocity.

velometer readings, the tubes will help you estimate the air velocity according to smoke behavior. Figure 12.4 illustrates how the smoke plume moves in different velocity airstreams.

- Studying spillage, turbulence around openings, and other factors that permit contaminants to escape from enclosures and canopies (Figure 12.3c). Since these phenomena occur randomly and in localized areas of enclosure openings, velometer readings may not identify them. Tracers are the techniques of choice in such cases.
- Determining the effect of crossdrafts on hood performance.
- Showing workers how ventilation hoods function and what effects distance, damper settings, and other factors have on hood performance.

Hood Velocity Measurements

Although smoke tube tests of hood performance are convenient and easy to perform, the results are qualitative. If the smoke flows into the hood

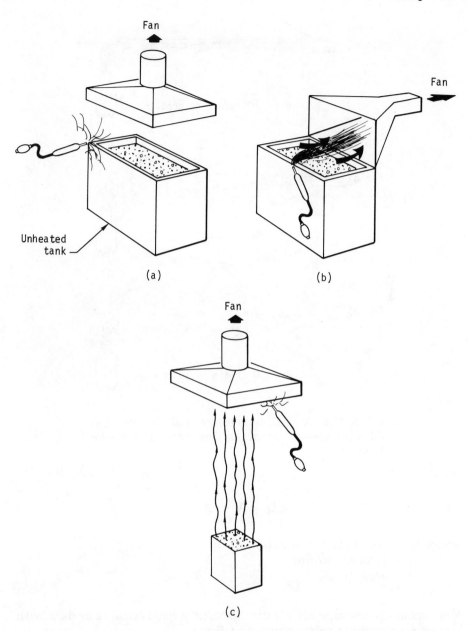

Figure 12.3 The use of smoke tubes to determine: (a) dispersion of contaminants; (b) range of capture velocity; and (c) spillage from canopy hoods or enclosures.

with reasonable velocity, the hood is judged to be adequate. Other tests are needed to give numerical air velocity readings. For example, volumetric airflow is often determined according to Equation 5.7:

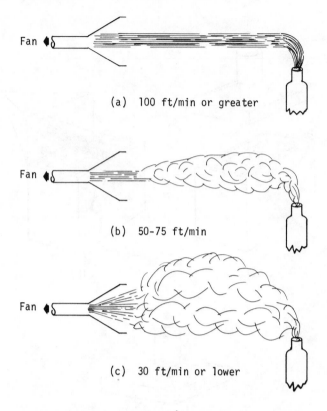

Fan

(a)   100 ft/min or greater

Fan

(b)   50-75 ft/min

Fan

(c)   30 ft/min or lower

Figure 12.4   Approximate behavior of a smoke jet in different velocity airstreams at location of smoke tube.

$$Q = V \times A$$

where Q = volumetric flow rate, ft³/min
V = velocity, ft/min
A = area, ft²

Measuring the average air velocity through a hood opening or duct with known area permits calculation of airflow.

Whenever air velocity readings are collected, it helps to record them either on a data sheet or a drawing of the ventilation system. Figure 12.1 shows a typical ventilation system and data sheet. By using the same format for subsequent tests, results can be compared to earlier tests to detect changes in system performance.

Air velocity can be measured with these instruments:

- Deflecting vane velometer (Figure 12.5) is the most common field velocity meter. It is available for measuring velocity from zero to 10,000 ft/min in a variety of hood openings. A modified Pitot tube and static pressure attachment (these tests are described later in this chapter) are also available. Typical applications include measuring capture velocity outside of capturing hoods, face velocity for enclosures (Figure 3.3), and slot velocities.
- Heated wire anemometer (Figure 12.6) works on the principle that the resistance of a heated wire changes with temperature variations. Air moving over the heated wire changes its temperature depending on the air velocity. The anemometer is calibrated to be read directly in feet per minute. Applications are similar to the deflecting vane velometer in Figure 12.5.
- Rotating vane anemometer (Figure 12.7) is useful in large openings such as doorways or large ventilated booths. It is not recommended for small areas where the instrument fills an appreciable portion of the opening.

Even though all of these instruments are factory calibrated before delivery, periodic calibration checks are recommended. In addition, cor-

Figure 12.5 Deflecting vane anemometer is popular for field velocity measurements. Some models have fittings for static pressure tests.

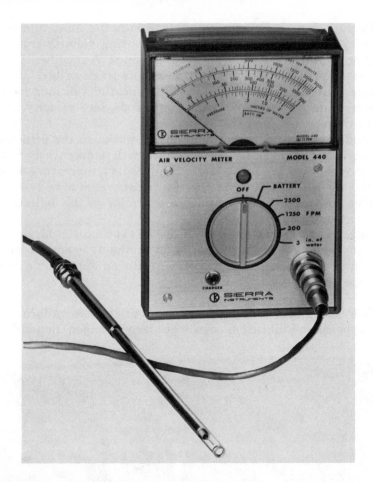

Figure 12.6   Heated wire anemometer permits direct reading of air velocity. *(Courtesy Sierra Instruments)*

rections to meter readings are needed for measurements on air with density that deviates significantly from 0.075 lb/ft³. Major causes for nonstandard air density are high or low temperatures, high humidity, or altitude. Manufacturers provide instructions on when and how to make these corrections.

Hood Static Pressure (Volume) Tests

Hood static pressure is the amount of suction in the duct near the hood. A certain hood static pressure is needed to draw the correct amount of air into the hood. If something happens to the ventilation system, such

Figure 12.7 Rotating vane anemometer is used to measure velocity in large openings such as spray booths. *(Courtesy Davis Instrument Manufacturing Co.)*

as a plugged duct or loose fan belt, the suction available at the hood declines and the hood draws less air. It is easily measured using a manometer (Figure 5.11).

Hood static pressure can also be considered the energy needed to overcome resistance to airflow into the hood. It is composed of the energy needed to accelerate room air to duct velocity plus the energy lost through turbulence and friction in the hood. Equation 5.5 for hood static pressure was derived in Chapter 5. Hood static pressure and volumetric airflow into the hood are related by Equation 5.12 for standard air (density = 0.075 lb/ft³):

$$Q = 4005 \ A \ C_e \ \sqrt{SP_h} \qquad (12.1)$$

where Q = volumetric airflow into hood, ft³/min
A = duct area, ft²
C_e = coefficient of entry for hood shape, dimensionless (from Table 5.1 or calculated from static pressure tests as described in Chapter 5.)
SP_h = hood static pressure, inches of water

Although the exact coefficient of entry may not be available for many hood shapes, the above equation is useful for calculating airflow after the hood static pressure has been measured.

Example: Calculate the volumetric airflow into the conveyor hood in Figure 5.11 assuming standard air density. Hood static pressure is 1.25 in. of water and the duct diameter is 6 in. For this hood C_e = 0.90.

Answer: Area = 0.2 ft² from Table 8.3.
From Equation 12.1:
$$Q = 4005 \, A \, C_e \, \sqrt{SP_h}$$
$$Q = 4005 \, (0.2) \, (0.9) \, \sqrt{1.25}$$
$$Q = 4005 \, (0.2) \, (0.9) \, (1.12) = 807 \text{ ft}^3/\text{min}$$

An easier way to establish a hood static pressure test program is to first determine airflows into hoods using velometer readings or a Pitot traverse across the ducts. Adjust airflow to each hood until it meets design criteria and then measure and record hood static pressure along with volumetric airflow. If subsequent hood static pressure readings show a decrease at a hood, the change in flow rate can be calculated using this equation:

$$\frac{Q}{Q_o} = \frac{\sqrt{SP_h}}{\sqrt{SP_o}} \qquad (12.2)$$

where Q = actual volumetric airflow rate, ft³/min
Q_o = original volumetric airflow rate, ft³/min
SP_h = actual hood static pressure, inches of water
SP_o = original hood static pressure, inches of water

Example: The hood static pressure at the welding bench in Figure 8.5 was 0.70 in. of water with airflow of 1050 ft³/min. Recent tests show the hood static pressure is now 0.55 in. of water. What is the current airflow into the hood?

Answer:
$$\frac{Q}{Q_o} = \frac{\sqrt{SP_h}}{\sqrt{SP_o}}$$
$$Q = Q_o \frac{\sqrt{SP_h}}{\sqrt{SP_o}}$$
$$Q = 1050 \text{ ft}^3/\text{min} \frac{\sqrt{.55}}{\sqrt{.70}} = 925 \text{ ft}^3/\text{min}$$

The decline in airflow could be caused by a loose fan belt, plugged duct, or other problem.

Measuring Hood Static Pressure

The principle of measuring hood static pressure is illustrated in Figure 5.11. A manometer or other pressure sensor is connected to a hole in the duct located about one duct diameter away from the hood. It is probably the easiest airflow test to perform. For ventilation systems where continuous flow measurements are needed, permanent installation of a manometer for hood static pressure readings is often the most convenient solution.

However, to give accurate tests, certain precautions must be followed. For example, the hole must be smooth on the duct interior wall. Any burrs or protruding edges cause turbulent airflow with increased velocity, which tends to lower hood static pressure readings. For this reason holes should be drilled rather than punched. A hole diameter of 1/16 to 1/8 in. is all that is necessary.[3] While a short length of metal tubing can be brazed onto the duct around the hole, another more convenient method entails holding the end of the rubber tubing against the duct. The hole does not have to be capped between tests if it is small enough. In large ventilation systems it may be worthwhile to number each static pressure hole so it can be located on the layout drawing and data sheet. If the holes are difficult to find in the field, paint a brightly colored ring around the test holes.[4]

Although a U-tube manometer is generally used for static pressure testing, mechanical pressure sensors may also be used. The manometer works on the principle that a column of water is displaced until the weight of liquid equals the pressure in the duct. As explained in Chapter 5, the pressure is expressed directly in "inches of water" column displaced, which is the distance between liquid levels in the two manometer legs. Mechanical pressure gauges, such as the magnahelic, bourdon tube gauge and the swinging vane anemometer, transmit pressure through mechanical linkages to a direct reading dial. They are convenient to use but should be calibrated periodically. Unlike the liquid-filled manometer, damage to the mechanical components may affect pressure readings.

So far, only static pressure tests at hoods have been discussed. Static pressure readings at the air cleaners, fan, and other duct locations are also important in assessing system performance and will be covered later in this chapter.

PITOT TUBE DUCT VELOCITY MEASUREMENTS

Although hood velocity or hood static pressure measurements are adequate for periodic tests, neither gives accurate determination of airflow through the system. The most accurate way to measure airflow in ducts is to perform a "Pitot traverse" of the duct. A Pitot tube is used to measure the air velocity at a number of points across the duct cross-section. Major applications of the Pitot traverse are for initial testing and balancing of new systems as well as for troubleshooting systems that are not performing adequately.

The Pitot tube (pronounced "pea-toe" after its inventor, Henri Pitot) is the standard device for measuring velocity pressure (and hence velocity) in ducts. For ventilation systems the tube shown in Figure 12.8 is used. It is made from two concentric tubes bent at a 90° angle and assembled so that the tip of the center tube is open and faces into the oncoming air stream. This tube measures the total pressure in the duct, which is the sum of the velocity pressure due to the moving air plus the static pressure exerted in all directions inside the duct. The outer tube is closed at the tip and has several small holes drilled through it perpendicular to the direction of airflow. The holes allow only the static pressure in the duct at the Pitot tube location to be detected. By connecting each of these tubes to a leg of a manometer, the static pressure is subtracted from the total pressure to yield velocity pressure in the duct (Figure 12.8). For standard air, velocity pressure can be converted to velocity by Equation 5.4:

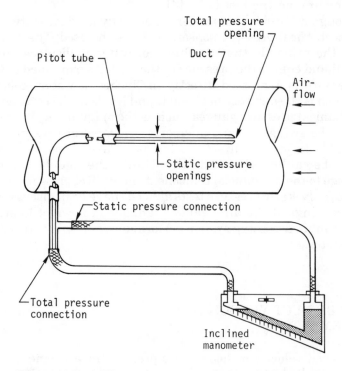

Figure 12.8   A Pitot tube connected to an inclined manometer is the primary method of measuring duct static, velocity, and total pressures. When connected as shown, the manometer indicates velocity pressure, which equals total pressure minus static pressure.

$$V = 4005 \sqrt{VP}$$

where V = air velocity, ft/min
       VP = velocity pressure, inches of water

Many manometers have both velocity pressure and velocity scales for direct reading of velocity under standard conditions. Table 8.4 also lists velocity and velocity pressure values.

The range of velocities that can be measured with a Pitot tube depends on the size of the tube and type of manometer. Below velocities of 1250–1500 ft/min (velocity pressures of about 0.10 to 0.14 in. of water), the U-tube manometer does not permit enough accuracy in pressure readings. The inclined manometer (Figure 12.9) expands the scale of pressure readings since it amplifies small vertical differences by a factor of about 10. The inclined manometer must be level for accurate readings and depending on design, high pressures may blow the liquid out of this type of manometer. However, it is essential for air velocities typical of industrial exhaust systems. Regardless of the manometer used, Pitot tube readings are inaccurate below about 300 ft/min velocity.

Pitot Traverse

A Pitot traverse involves measuring the velocity at a number of points across the duct area since velocity distribution is not uniform within the duct. As a rule of thumb the average duct velocity is about 90% of the centerline velocity; however, the accuracy of this rule of thumb is insufficient for many tests. Whenever turbulent or stratified airflow in the duct is suspected, a traverse is necessary. The number and location of measuring points within the duct depend on the duct size and shape.

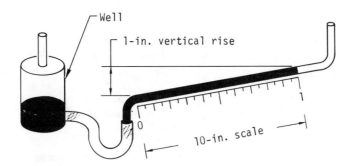

Figure 12.9   To increase precision, an inclined manometer amplifies slight vertical movement by a factor of 10. (Source: Reference 5)

The idea is to divide the duct into enough zones of equal area to give accurate results according to these guidelines:[5]

- For round ducts two traverses at right angles should be made. For ducts 6 in. in diameter or smaller, make two 6-point traverses. For ducts 6 to 48 in., make at least two 10-point traverses. Above 48 in. or for smaller ducts where large velocity variations are suspected, make two 20-point traverses. The locations of the measuring points are selected in such a way that the duct is divided into equal annular areas (Figure 12.10), not equidistant points along the duct diameter. Tables 12.2 and 12.3 show the locations for 6- and 10-point traverses for duct diameters up to 48 in.
- For rectangular ducts, the cross-section is divided into equal areas and a reading is taken at the center of each area (Figure 12.11). At least 16 readings should be taken, but the distance between measuring points should not exceed 6 in.

Regardless of duct shape the best place to perform a traverse is at least 7.5 duct diameters downstream from any major disturbance to smooth airflow such as dampers or elbows. If you must traverse closer

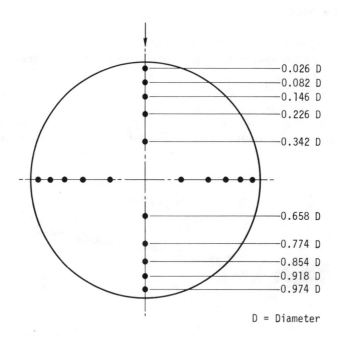

Figure 12.10   Measuring locations for a 10-point Pitot traverse in a round duct. These points divide the duct into equal area annular rings.

Table 12.2   Round Duct 6-Point Pitot Traverse Locations

Duct Diameter (inches)	*Distance from Wall of Round Duct to Point of Reading^a (inches)* *Traverse Point*					
	1	*2*	*3*	*4*	*5*	*6*
3	⅛	½	⅞	2⅛	2½	2⅞
3½	⅛	½	1	2½	3	3⅜
4	⅛	⅝	1⅛	2⅞	3⅜	3⅞
4½	¼	⅝	1⅜	3⅛	3⅞	4¼
5	¼	¾	1½	3½	4¼	4¾
5½	¼	¾	1⅝	3⅞	4¾	5¼
6	¼	⅞	1¾	4¼	5¼	5¾

^aTo the nearest ⅛ in.

than 7.5 diameters, make another traverse at a second location and compare results. If the calculated volumetric flow rates are within 10% of each other, the results are satisfactory.

Along with velocity pressure readings, also record the temperature of air in the duct and the amount of moisture. The moisture is usually determined with wet bulb and dry bulb thermometer readings. The amount of moisture is found using a psychometric chart (Figure 12.12). If moisture exceeds about 0.2 pounds of water per pound of air or if air temperature is outside the 40°F to 100°F range, or if altitude exceeds

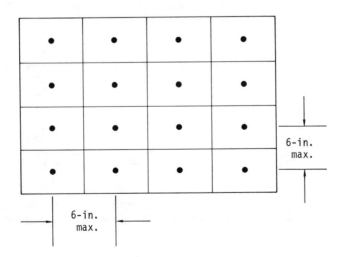

Figure 12.11 In a rectangular duct take at least 16 Pitot readings, but the measuring points should not be more than 6 inches apart.

Table 12.3   Round Duct 10-Point Pitot Traverse Locations

Duct Diameter (inches)	Distance from Wall of Round Duct to Point of Reading[a] (inches) Traverse Point									
	1	2	3	4	5	6	7	8	9	10
4	⅛	⅜	⅝	⅞	1⅜	2⅝	3⅛	3⅜	3⅝	3⅞
4½	⅛	⅜	⅝	1	1½	3	3½	3⅞	4⅛	4⅜
5	⅛	⅜	¾	1⅛	1¾	3¼	3⅞	4¼	4⅝	4⅞
5½	⅛	½	¾	1¼	1⅞	3⅝	4¼	4¾	5	5⅜
6	⅛	½	⅞	1⅜	2	4	4⅝	5⅛	5½	5⅞
7	⅛	⅝	1	1⅝	2⅜	4⅝	5⅜	6	6⅜	6⅞
8	¼	⅝	1⅛	1¾	2¾	5¼	6¼	6⅞	7⅜	7¾
9	¼	¾	1¼	2	3⅛	5⅞	7	7¾	8¼	8¾
10	¼	⅞	1½	2¼	3⅜	6⅝	7¾	8½	9⅛	9¾
11	¼	⅞	1⅝	2½	3¾	7¼	8½	9⅜	10⅛	10¾
12	⅜	1	1¾	2¾	4⅛	7⅞	9¼	10¼	11	11⅝
13	⅜	1	1⅞	2⅞	4½	8½	10⅛	11⅛	12	12⅝
14	⅜	1⅛	2	3⅛	4¾	9¼	10⅞	12	12⅞	13⅝
15	⅜	1¼	2¼	3⅜	5⅛	9⅞	11⅝	12¾	13¾	14⅝
16	⅜	1¼	2⅜	3⅝	5½	10½	12⅜	13⅝	14¾	15⅝
17	½	1⅜	2½	3⅞	5¾	11¼	13⅛	14½	15⅝	16½
18	½	1½	2⅝	4⅛	6⅛	11⅞	13⅞	15⅜	16½	17½
19	½	1½	2¾	4¼	6½	12½	14¾	16¼	17½	18½
20	½	1⅝	2⅞	4½	6⅞	13⅛	15½	17⅛	18⅜	19½
22	⅝	1¾	3¼	5	7½	14½	17	18¾	20¼	21⅜
24	⅝	2	3½	5½	8¼	15¾	18½	20½	22	23⅜
26	⅝	2⅛	3¾	5⅞	8⅞	17⅛	20⅛	22¼	23⅞	25⅜
28	¾	2¼	4⅛	6⅜	9⅝	18⅜	21⅝	23⅞	25¾	27¼
30	¾	2½	4⅜	6¾	10¼	19¾	23¼	25⅝	27½	29¼
32	⅞	2⅝	4⅝	7¼	11	21	24¾	27⅜	29⅜	31⅛
34	⅞	2¾	5	7¾	11⅝	23⅜	26¼	29	31¼	33⅛
36	1	3	5¼	8⅛	12⅜	23⅝	27⅞	30¾	33	35
38	1	3⅛	5½	8⅝	13	25	29⅜	32½	34⅞	37
40	1	3¼	5⅞	9	13⅝	26⅜	31	34⅛	36¾	39
42	1⅛	3⅜	6⅛	9½	14⅜	27⅝	32½	35⅞	38⅝	40⅞
44	1⅛	3⅝	6⅜	10	15	29	34	37⅝	40⅜	42⅞
46	1¼	3¾	6¾	10⅜	15¾	30¼	35⅝	39¼	42¼	44¾
48	1¼	4	7	10⅞	16⅜	31⅝	37⅛	41	44	46¾

[a]To the nearest ⅛ in.

±1000 ft relative to sea level, density corrections are needed to velocity pressure readings. For temperature and atmospheric pressure the density correction is:

$$\text{Density}_{actual} = \frac{0.075 \text{ lbs}}{\text{ft}^3}\left[\frac{530°F}{(460 + t)°F}\right]$$
$$\times \left[\frac{\text{Barometric Pressure, in. of Hg}}{29.92}\right] \quad (12.3)$$

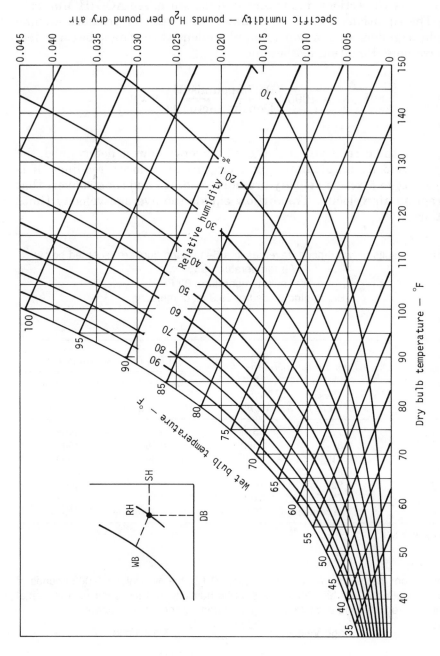

Figure 12.12 A psychometric chart is used to find relative humidity or specific humidity from wet bulb and dry bulb temperature readings.

where t = temperature, °F

For density corrections due to elevated moisture, see ACGIH *Manual.*
    The equations derived in Chapter 5 for velocity pressure assume standard air density. For nonstandard air density they must be corrected. For example, Equation 5.4 becomes:

$$V = 4005 \left(\frac{0.075 \text{ pounds/ft}^3}{\text{Density (actual)}} \right) \sqrt{VP}$$

    Once the correct velocity at each traverse point is determined, the average velocity is calculated by averaging all velocity readings; it is not correct to average velocity pressure readings and then convert that average to air velocity. The airflow equals the average velocity times duct area.

Example: Results of two 6-point traverses in a 6-in.-round duct are recorded below.
Air temperature: 85°F
Wet bulb temperature: 60°F
Barometric pressure: 29.9 in. of mercury

	Pitot Traverse Number 1		Pitot Traverse Number 2 (perpendicular to Traverse 1)	
Point	Velocity Pressure (in. of water)	Velocity* (ft/min)	Velocity Pressure (in. of water)	Velocity* (ft/min)
1	0.77	3514	0.75	3468
2	0.80	3582	0.78	3537
3	0.91	3821	0.90	3800
4	0.89	3779	0.87	3729
5	0.82	3625	0.81	3604
6	0.75	3468	0.73	3422
	Average:	3631.5	Average:	3593.3

*From Table 8.3.

Answer: From the psychometric chart (Figure 12.12), the air contains 0.005 pounds of water per pound of dry air. Since this specific humidity and the other environmental conditions are within the acceptable range, no density correction is needed.

$$\text{Average Velocity} = \frac{3632 + 3593}{2} = 3613 \text{ ft/min}$$

From Table 8.3, the area of a 6-in.-diameter duct is 0.1964 ft².
Q = V × A = 3613 ft/min × 0.1964 ft² = 710 ft³/min

Pitot Tube Detects Spinning Flow

Fan performance can be seriously affected by spinning airflow at the fan inlet. Manufacturers' fan ratings assume that the air entering the fan is moving in a straight line, that there is no spin component.

The only way to detect spinning flow is with a Pitot tube. Insert the tube into the duct near the fan (connected as shown in Figure 12.8) and twist it slowly up and down, and back and forth at different points within the duct. The maximum velocity pressure reading occurs when the Pitot tube is facing directly into the air stream. If maximum readings are found with the Pitot tube pointed slightly upward on one side of the duct and downward on the other side of the duct, the approximate angle of spin can be estimated.[6] Retest the system after flow straighteners or elbow turning vanes have been installed to see whether the condition has been corrected.

SYSTEM STATIC PRESSURE TESTS

The use of hood static pressure tests to measure airflow into hoods was explained earlier. Static pressure tests at other system locations are also valuable to measure system performance and diagnose malfunctions. The static pressure at any point is the potential energy available at that location for overcoming friction, turbulence, and other causes of pressure loss in the system upstream (toward the hoods) from that point. Static pressure tests on either side of filters to measure pressure loss through the unit are the standard way to determine when the filter is clogged and needs cleaning or replacing. Except for pressure readings across filters or similar units, the value of static pressure readings comes from comparing present readings with static pressure values recorded when the system was operating properly.

Other than at each hood, typical locations to measure static pressure are: at entries into main ducts, on each side of air cleaners, on each side of the fan, and at several points on long ducts. Initial and periodic static pressure readings should be recorded on a data sheet. Usually differences between readings at the same locations that exceed 10% should be investigated to determine the reason for static pressure drop.

The testing technique is similar to hood static pressure tests. Small diameter holes (1/16 to 1/8 in.) are drilled through the duct walls so no burrs protrude into the flowing airstream. The holes should be at least 7.5 duct diameters downstream from any disturbance such as an elbow, damper, or branch duct entry. If this is not possible, then four holes should be drilled 90° apart around the duct and static pressure values averaged. The end of a rubber tube attached to a U-tube manometer or other pressure sensor is pressed against the duct around the hole; the static pressure is read as "inches of water" (Figure 5.11). If accurate pressure readings are desired, the static pressure arm of a Pitot tube connected to one side of a manometer is more accurate than a pressure tap in the duct wall.

Here is an example of how static pressure tests can help diagnose ventilation system problems. A ventilation system (Figure 12.13) consists of six hoods and branch ducts, a main duct, an air cleaner, and a fan. Over several years the following problems were identified:[7]

- The hood static pressures are low in all hoods (Figure 12.13a). Further tests show that static pressure on the hood side of the air cleaner is low, while it is higher than usual between the air cleaner and fan. These readings indicate that the air cleaner is causing too much resistance and should be cleaned to restore proper performance.
- The hood static pressures are low in all hoods (Figure 12.13b) but the readings on both sides of the air cleaner are above normal. This indicates that the main duct or branch duct is partially plugged.
- The hood static pressure readings are low (Figure 12.13c) as is the static pressure on both sides of the air cleaner. This indicates that the fan is not working properly, the discharge stack is plugged, or there is a loose duct joint.
- All static pressure readings are normal except one hood static pressure, which is too low (Figure 12.13d). This indicates that the branch duct is plugged between the test point and the main duct.
- All static pressure readings are near normal except one hood static pressure reading, which is increased (Figure 12.13e). This means that a blockage exists between the hood opening and the static pressure test point.
- All static pressure readings are near normal except the hood static pressure readings in two adjacent hoods, which are decreased (Figure 12.13f). This indicates that the main duct is plugged near the two hoods with lower airflow.

All of these difficulties were quickly solved once the source of trouble was recognized. In each case a blockage caused a higher static pressure reading on the fan side of the blockage because the blockage prevented the suction (static pressure) from being converted into velocity pressure.

TESTING FAN PERFORMANCE

Fans do not need much attention except for periodic maintenance as long as the system is operating properly. If power consumption is higher than it should be, if decreased airflow through hoods is attributed to poor fan

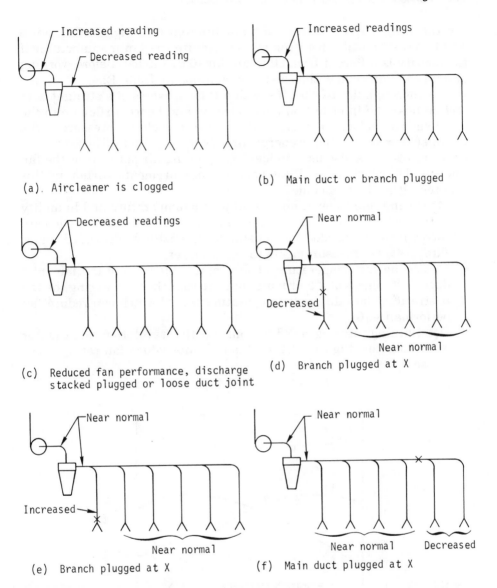

(a) Aircleaner is clogged

(b) Main duct or branch plugged

(c) Reduced fan performance, discharge stacked plugged or loose duct joint

(d) Branch plugged at X

(e) Branch plugged at X

(f) Main duct plugged at X

Figure 12.13 Ventilation system problems are diagnosed by comparing current static pressure readings with earlier tests.

performance, or if additions to the ventilation system are being considered, tests of fan performance may be helpful. To get the most value from fan tests, you need a fan rating curve from the manufacturer for your model and size fan at the same air density and speed that your fan is operating. Usually the manufacturer supplies a single drawing with several rating curves corresponding to different rotating speeds. Figure 12.14 shows typical fan rating curves. Advise the fan manufacturer if the density is different from standard air (0.075 lb/ft³). Before working around the fan, review the safety precautions in Table 12.4.

A thorough fan test begins with a Pitot traverse of a straight duct section near the inlet to find duct velocity (and hence airflow) into the fan. Fan inlet velocity is important because the velocity pressure at the fan inlet represents kinetic energy already in the system. These velocity pressure tests are also used to identify spinning air patterns at the fan inlet as described under Pitot velocity measurements earlier in this chapter. Other tests include:

- Measuring plant elevation as well as air temperature and humidity in the system. From these data you can determine whether air density corrections are needed for nonstandard density as discussed under Pitot traverse measurements in this chapter.
- Measuring air temperature at four places 90° around the fan inlet duct. Differences of 2°F or more indicate that the air entering the fan is stratified into different temperature zones, which can reduce fan performance.[6]
- Determining rotating speed (rev/min) with a revolution rate counter or tachometer (Figure 12.15). This indicates which fan rating curve to use.

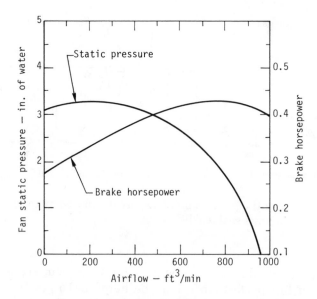

Figure 12.14 Typical fan rating curves relating fan static pressure and brake horsepower to volumetric airflow for a single rotating speed.

Table 12.4 Safety Precautions for Fan Testing

Air blasts from fan outlets or exhaust stacks can cause injury.
Do not open fan access doors while the system is running. On the fan outlet side, the doors can slam open with explosive force.
Close all access doors before starting the fan motor. An open door on the suction side of the fan can overload the motor during starting and can also create a suction hazard.
If fan belt or shaft guards must be removed to measure rotating speed, beware of entangling long hair, ties, and so forth, in rotating machinery.
Whenever the fan is operated with guards removed or ducts disconnected, clear the area of unauthorized personnel and post signs or guards to keep bystanders away.

Source: Reference [6].

Figure 12.15 Photo-tachometer for measuring fan rotating speed. *(Courtesy Pioneer Electric and Research Corp.)*

- Measuring fan inlet and outlet static pressures with either a static tap in the duct wall (Figure 9.3) or the static pressure arm of a Pitot tube and a manometer. The static readings must be corrected for nonstandard air density. With these readings plus the fan inlet velocity you can calculate fan static pressure according to Equation 8.2:

$$FSP = |SP_{inlet}| + |SP_{outlet}| - VP_{inlet}$$

where FSP = fan static pressure, inches of water
  SP = static pressure, inches of water
  VP = velocity pressure, inches of water
  inlet, outlet = fan inlet and outlet

Example: A Pitot traverse shows that duct velocity in the 6-in. diameter fan inlet is 3613 ft/min. If static pressure readings are 2.25 in. of water at the fan inlet and 0.55 in. of water at the outlet, what is the fan static pressure?

Answer: From Table 8.4 the velocity pressure is 0.81 in. of water for a velocity of 3613 ft/min.

$$FSP = 2.25 + 0.55 - 0.81 = 1.99 \text{ in. of water}$$

- Determining the volts and amps supplied to the fan motor. A clamp-on voltmeter-ammeter (Figure 12.16) is convenient to use. If you can obtain motor efficiency and power factor from the manufacturer, you can calculate motor horsepower for a three-phase motor:

Figure 12.16 Clamp on ammeter-voltmeter for measuring fan operating horsepower. *(Courtesy Arrow-M Corp.)*

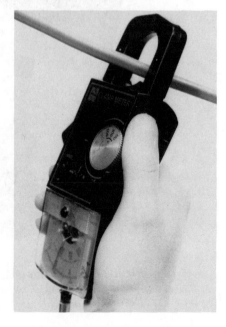

$$BHP = \sqrt{\frac{3 \times \text{Volts} \times \text{Amps} \times \text{PF} \times \text{E}}{746}} \qquad (12.4)$$

where BHP = brake horsepower
  PF = motor power factor
  E = motor efficiency

The following equation can be used to estimate fan horsepower, if the volumetric airflow and total pressure in the system are known:

$$BHP = \frac{Q \times TP}{6356 \times ME} \qquad (12.5)$$

where Q = volumetric flow rate, ft³/min
  TP = fan total pressure, inches of water
  ME = fan mechanical efficiency (usually 0.50 to 0.65)

The fan total pressure is the total energy (static and velocity pressures) added to the system by the fan:

$$TP = |SP_{inlet}| + VP_{outlet} + |SP_{outlet}| - VP_{inlet} \qquad (12.6)$$

$$\text{Since } FSP = |SP_{inlet}| + |SP_{outlet}| - VP_{inlet}$$

Equation 12.6 becomes

$$TP = FSP + VP_{outlet} \qquad (12.7)$$

where $SP_{inlet, outlet}$ = static pressure at fan inlet and outlet, inches of water
  VP = velocity pressure, inches of water
  FSP = fan static pressure, inches of water

Fan horsepower must also be adjusted for nonstandard air density if environmental conditions warrant.

  If motor efficiency and power factor data are not available, at least see that the motor is operating within its nameplate amperage rating.

• Locating the fan static pressure, air volume, and horsepower operating points (Figure 12.17) by using the fan characteristic curve for the fan's rotating speed. Fan output can be plotted three ways on the curve: from the fan static pressure curve, the motor horsepower curve,

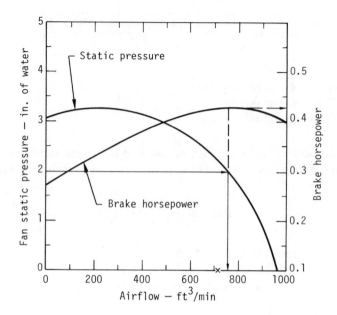

Figure 12.17 Fan rating curves from Figure 12.14 showing how airflow and horsepower are estimated from fan static pressure. The (X) on the horizontal axis marks the airflow calculated from a Pitot traverse test.

and the Pitot traverse at the fan inlet.[6] They should all agree fairly closely. If differences are great, reduced fan performance may be a problem. Review the fundamentals of fan operation in Chapter 9 and the solutions to ventilation problems in Chapter 13.

Example: Using the previous example problem, plot and compare the volumetric airflows from the fan static pressure calculation and the Pitot tube survey. Also calculate brake horsepower. Use the fan curve in Figure 12.17. Assume that fan outlet velocity is the same as fan inlet velocity.

Answer: The average duct velocity in the 6-in.-diameter duct was 3613 ft/min according to the Pitot traverse. From Table 8.3 the area of a 6-in. duct is 0.2 ft².

$$Q = V \times A = 3613 \text{ ft/min} \times 0.2 \text{ ft}^2 = 723 \text{ ft}^3/\text{min}$$

Mark this on the horizontal axis of Figure 12.17 with an "x".

Next enter Figure 12.17 at 1.99 in. of water on the vertical scale, since this is the fan static pressure calculated previously. Find the volumetric flow rate corresponding to this value by using the fan static pressure rating curve (760 ft³/min).

Since both airflow values are close, assume the difference is due to test errors. Average the two to estimate actual flow.

$$Q = \frac{723 + 760}{2} = 742 \text{ ft}^3/\text{min}$$

Estimate brake horsepower using Equation 12.5. Assume a mechanical efficiency of 0.60, since the actual value is unknown:

$$BHP = \frac{Q \times TP}{6356 \times ME}$$

From Equation 12.7:

$$TP = FSP + VP_{outlet}$$
$$TP = 1.99 + 0.81 = 2.8 \text{ in. of water}$$
$$BHP = \frac{742 \times 2.8}{6356 \times .6} = 0.54 \text{ horsepower}$$

This is higher than the 0.42 brake horsepower read from Figure 12.17, which probably reflects inaccuracies in assuming a fan mechanical efficiency of 0.60 when the actual value is unknown. However, it does illustrate the use of Equation 12.5 in estimating horsepower when the fan brake horsepower curve is not available.

PERIODIC MAINTENANCE CHECKS

Periodic maintenance is the best way to assure that the system will continue to give the health protection it was installed to provide. Periodic maintenance and inspection includes: lubrication of fan, motor, and drive; a check of fan belt for proper tension; inspection of air cleaner internals for visible damage; inspection of fan wheel and housing for wear and dust buildup; and a check of dust systems for partially plugged ducts. The size and location of inspection ports depend on duct diameter.

SUMMARY

Testing to assess and document performance is important for every local exhaust system. A new system should be tested to assure that it meets design specifications; an existing system requires periodic checks to establish that performance has not declined.

Where proper airflow into hoods is the major concern, smoke-tube tracer studies or hood velocity measurements often are sufficient. Hood static-pressure measurements are easy to perform and reveal any drop in the amount of suction at each hood that is available to draw contaminated air into the system. Factors reducing suction that are revealed by hood static-pressure tests include plugged ducts, clogged air cleaner, or fan mechanical problems.

Thorough studies of system and fan conditions may involve Pitot-tube duct velocity measurements, system-wide static pressure tests, or fan evaluations such as rotating speed and power consumption. These tests help to diagnose system problems and determine if an existing system can be expanded simply by adding additional hoods.

REFERENCES

1. ACGIH Committee on Industrial Ventilation. *Industrial Ventilation—A Manual of Recommended Practice,* 17th Ed. (Lansing, Michigan: American Conference of Governmental Industrial Hygienists, 1982).
2. Hemeon, W.C.L. *Plant and Process Ventilation* (New York, New York: Industrial Press, Inc., 1963).
3. American National Standard Z9.2—1971. "Fundamentals Governing the Design and Operation of Local Exhaust Systems" (New York, New York: American National Standards Institute, 1972).
4. National Safety Council. "Checking Performance of Local Exhaust Systems," Data Sheet 438 (Chicago, Illinois: National Safety Council, 1963).
5. ACGIH Committee on Industrial Ventilation. "Ventilation System Testing" (Lansing, Michigan: American Conference of Governmental Industrial Hygienists, 1971).
6. Trickler, C.J. "Field Testing of Fan Systems," Engineering Letter E-3 (Chicago, Illinois: The New York Blower Company).
7. Tennessee Department of Health. "Tennessee Industrial Hygiene News" (1950).

CHAPTER 13

Solving Ventilation System Problems

Many ventilation systems do not work properly. Difficulties generally fall into one of these three categories:

- Airflow into hoods does not meet design specifications or OSHA standards at one or more hoods.
- Airborne levels of contaminants are too high even though the system seems to be operating properly.
- Noise from the ventilating system is so high that it creates either a hazard to hearing or a nuisance within the plant.

If a ventilating system is not working correctly, then one or more principles governing local exhaust has been violated. If poor airflow is the problem, look for solutions in Chapters 5, 8, and 9 on airflow principles, system design, and fan selection. If the airborne contaminant levels are too high, then refer to the discussion of hoods in Chapter 6. Guidelines for minimizing ventilation system noise are included in Chapter 9 on fans. Your job is to find out why the system is not operating correctly and how you can fix it.

FIRST STEP: DIAGNOSIS

Diagnose the problem by thoroughly inspecting the system and by taking pressure and velocity readings as described in Chapter 12. The visual inspection will reveal closed dampers, open inspection ports, damaged hoods and ducts, and other common reasons for poor performance. The static pressure measurements at hoods, elbows, and on both sides of air cleaners will show the contribution of each to the overall pressure drop in the system. Static pressure measurements on both sides of the fan show how much pressure the fan is adding. If you know that the system ever operated correctly, try to compare current pressure readings with previous data. If no earlier data are available, try estimating what the pressure drop readings should be from the design tables in Chapter 8 or in the ACGIH *Industrial Ventilation Manual*.[1] For new systems, pressure readings will help detect installation mistakes or blockages due to construction debris.

Static pressure is an important parameter, since it is the suction that pulls air in through the hoods and then through the ducts to the fan. In order for the system to operate, the fan must move the correct amount of air as well as develop the correct static pressure. Thus, the energy that the fan adds to the system is expressed as "fan static pressure." Since the fan's volumetric output (ft³/min) decreases as static pressure (resistance) in the system increases (Figure 13.1), any change in the system that increases resistance or static pressure loss will cause a reduction in airflow.

Although static pressure readings are important, they cannot be used to detect some kinds of malfunctions. A common reason for too little airflow is a poor fan inlet connection (Figure 9.8) that causes spinning

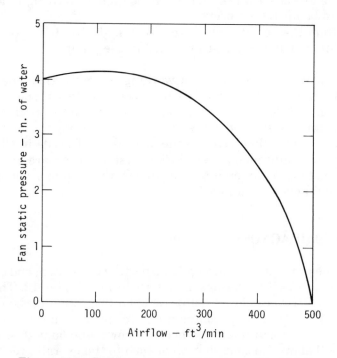

Figure 13.1 Static pressure rating curve for a backward inclined fan shows how the fan's volumetric output varies with system resistance.

airflow or uneven flow patterns into the fan. These problems prevent the fan from doing its maximum rated work on the moving air and they reduce fan volumetric output. Pressure readings before and after the fan will not reveal the problem, since the fan is generating the correct amount of pressure. Pitot tube tests to detect spinning airflow are covered in Chapter 12.[2]

Hood performance can be assessed using velocity measurements and smoke tube tracer studies. Velocity tests show how the air flowing into the hood is distributed. Achieving the specified volumetric airflow through each hood is not enough; the velocity must be distributed over the area of contaminant generation in order to capture or contain the airborne material. The smoke tube tests will help show how and where contaminants are escaping from the hoods into the workroom.

Maintenance Problems

Problems such as loose fan belts, dirty filters, plugged ducts, or dirty fan blades can be easily fixed. Remember that centrifugal fans will move some air even when they are rotating backwards—so check fan rotation after electrical work on the fan motor. Keep in mind that it is a lot easier to identify maintenance problems if static pressure and velocity measurements from the correctly operating system are available for comparison with current readings.

Hood Problems

Hood problems cause excessively high contaminant levels in the workroom. Situations where airflow into hoods does not meet design criteria are covered under fan and system problems in Chapters 8 and 9. Problems occur when hoods are not selected and designed with both the industrial process and contaminants in mind. Remember that there are three different hood types (Figure 6.7):

- Capturing hoods that reach out to draw in contaminants from outside of the hood. These hoods function by developing the needed capture velocity at the point where contaminants are released.
- Enclosures that surround the contaminant source and contain contaminants released inside the enclosures. An inward airflow through openings in the enclosure prevents contaminant escape.
- Receiving hoods that catch contaminants propelled toward the hood by thermal currents or centrifugal action.

Capturing Hoods

The capturing hood is often the most convenient type of hood to use since it can be located adjacent to the process without surrounding it. However, in many ways it is also the most difficult type of hood to use. Look for these problems with capturing hoods (Figure 13.2):

- Hoods with insufficient airflow being expected to reach outside the hood to capture and draw in contaminants. Capture hoods usually need a lot more airflow than other hood types (Figure 13.2a).

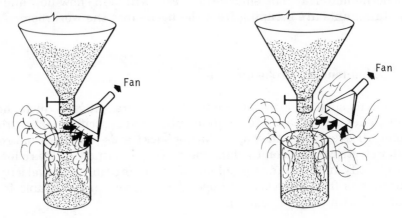

(a) Insufficient airflow at most distant point of contaminant release

(b) Hood located too far from source

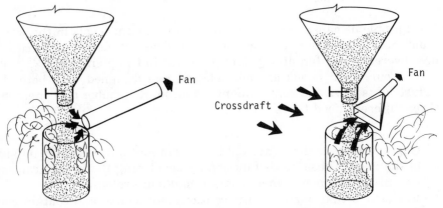

(c) Poor velocity distribution across zone of release

(d) Crossdrafts reduce capture efficiency

Figure 13.2 Common reasons why capturing hoods do not work properly.

- Capturing hoods located too far from the source of contaminants (Figure 13.2b). Velocity drops off severely with increased distance from the hood face. Often several inches is the maximum effective reach of a capturing hood.
- Capturing hoods with poor velocity distribution over the zone of contaminant release (Figure 13.2c). Formulas for calculating capture velocity such as Equation 6.1 refer to the centerline velocity only. Centerline velocity refers to the velocity along a line extending out from the center of the hood face. If the velocity distribution across the hood face is unsatisfactory, the necessary capture velocity may not be developed at some locations where contaminants are being generated.
- Capturing hoods that have not been designed according to the null point theory of capture (Figure 6.20).[3] The null point theory says that contaminants released with initial velocity away from the hood will not be captured until they slow down enough for the capture velocity to control them. This means that the capture velocity must be developed not just where the contaminants are released but at the point where the initial dispersive forces of the contaminants are expended.
- Capturing hoods that are not designed to overcome thermal rise, crossdrafts, and other factors tending to disperse contaminants in the workroom (Figure 13.2d). The capture velocity criteria in Table 6.5 refer to fairly optimum conditions. Figure 6.15 shows that a crossdraft of 60 ft/min can reduce capturing hood efficiency by 50%.

The solutions to most of these problems will be evident once the cause is determined.

Enclosures

Enclosures usually work well if the face velocity is sufficient to contain contaminants and if turbulent forces that upset inward airflow into the hood are minimized. Here are typical problems (Figure 13.3):

- Poor velocity distribution across hood openings results in an inward face velocity that is too low to control contaminants at some points (Figure 13.3a). Refer to the description of laboratory hoods in Chapter 6 for the importance of air distribution.
- The enclosure is not large enough to surround the zone where the initial dispersive forces of the contaminants are spent (Figure 13.3b). Face velocities of 100 to 150 ft/min are not adequate to slow down and control contaminants traveling out of the hood with high velocity. If the process generates contaminants with initial velocity, they may escape from the hood.

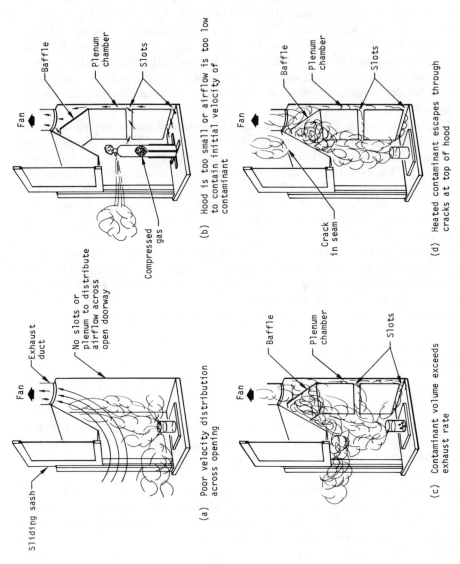

Figure 13.3 Common reasons why enclosures do not work properly.

- The volume of contaminants exceeds the exhaust volume from the hood. A type of "overflow" results (Figure 13.3c).
- Heated contaminants escape from cracks or other openings in the top of the enclosure (Figure 13.3d) due to their thermal head.[3]

Receiving Hoods

Receiving hoods are not used very often for contaminant control systems, since they cannot always control gases, vapors, and fine particulates that cause inhalation hazards to workers. If a receiving hood is being used for any purpose other than use over a heated process where the rising heated air carries contaminants into the hood, use a smoke tube or other technique to look for contaminants escaping into the workroom. If the application is not a good one for a receiving hood, replace it with an enclosure or capturing hood.

Solving Hood Problems

Hood problems can be solved by redesigning or relocating the hood or, in some cases, increasing the airflow. An enclosure usually gives the best contaminant control with the lowest airflow requirement. Baffles and slots help to give good air distribution. Protecting the hood from crossdrafts or other dispersive forces, either by relocating the process or shielding the hood, will help the hood work as it should.

Increasing the airflow into only a few hoods in a multiple-hood system may be difficult. During design and after installation the system was balanced to give the design airflow through each hood. Since increasing the flow in some hoods may decrease the airflow through others, the whole system should be considered when increasing airflow. Usually the solution is to increase airflow through the entire system to meet minimum airflow criteria in all hoods as discussed later.

Too many hoods in a system can also be a problem if some were added to the system without proper design (Figure 13.4).[4] If the fan has sufficient excess capacity and if the new hood and branch duct are designed to match the static pressure in the existing ducts, small volume hoods can often be safely added. Unfortunately this is usually not the case.

Example: A glove box (Figure 10.7) exhausting 30 ft^3/min is to be added to an existing ventilation system (Figure 13.5). Assuming that the fan can handle the additional 30 ft^3/min without a measurable drop in fan static pressure output, where should the glove box be connected if the static pressure drop through the glove box and connecting branch duct is 3.0 in. of water?

Answer: As long as the fan can handle the additional 30 ft^3/min air volume with no problem, the challenge is to find where to connect the new duct into the existing system.

Figure 13.4 Additional hoods and ducts added to an existing system often upset airflow throughout the entire system.

Static pressure readings obtained with a manometer (Figure 5.6a) are listed in Figure 13.5. Any point between point E and the filter will be satisfactory since the static pressure in the system exceeds 3.0 in. of water. If the glove box was connected before point E, there would be too little suction in the main duct to draw 30 ft³/min through the glove box.

Adding hoods by this procedure will work in systems designed with dampers for balance. In systems designed without dampers, adding new hoods and ducts may upset the whole system balance and require dampers in the branch ducts.

Duct Problems

Ducts have two major functions in an exhaust ventilation system. Their primary function is to carry air between the hoods, air cleaner, fan, and discharge stack. Their second function is to provide pressure loss, or resistance, in multiple-hood systems. This results in the proper distribution of airflow among the hoods.

Many duct systems have features that cause unnecessary pressure drop, which reduces fan output. Narrow ducts, tight elbows, and perpendicular branch entries (Figure 13.6) are all culprits. Some hood de-

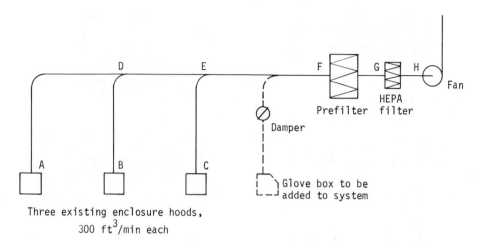

Figure 13.5 caption:

Three existing enclosure hoods,
300 ft³/min each

Location	Static pressure, in. of water
A	1.25
B	1.25
C	1.25
D	2.65
E	3.10
F	3.75
G	5.60
H	8.70

Figure 13.5 A new hood or enclosure can be added to an existing system only if the fan can handle the extra volume. Connect the new duct where the suction is great enough to draw the needed airflow.

signs also cause too much turbulence, which reduces airflow. Narrow slot openings or orifices and straight duct takeoffs (Figure 13.7) are typical high-pressure loss features in hoods.

For single-hood systems, every source of pressure drop that you can identify and reduce will increase airflow through the system. The amount of increase can be estimated from the rating curve for the fan installed in the system.

Example: A slot hood system on an open surface tank (Figure 13.8) has a fan static pressure requirement of 3.5 in. of water at 300 ft³/min airflow.

The hood contributes 1.8 in. of water resistance. Investigation revealed that the slots could be opened without reducing hood performance. Opening the slots reduced

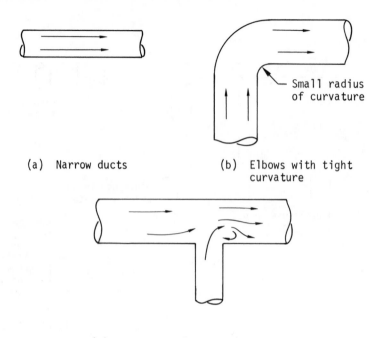

(a)   Narrow ducts

(b)   Elbows with tight
      curvature

(c)   Perpendicular duct junctions

Figure 13.6   Three ways that poor duct design reduces airflow
through the system.

the hood resistance to 0.5 in. of water, thus changing the system fan static pressure
requirement from 3.5 to 2.2 in. of water. Use the fan curve in Figure 13.1 to estimate
the new airflow through the system.

Answer: Reducing the system resistance will increase the amount of air the fan will pull
through the system.

   The new airflow can be calculated by a trial-and-error procedure based on techniques in Chapter 8. It can be estimated from the fan rating curve (Figure 13.1). The
fan will move about 400 ft³/min against 2.2 in. of water fan static pressure.

   The final volume will be slightly below 400 ft³/min, however, since the duct friction
will increase slightly as the airflow increases. Thus the final system static pressure will
exceed 2.2 in. of water at 400 ft³/min.

Path of Greatest Resistance

In multiple-hood systems, reducing the pressure loss through individual
hoods and branch ducts will help only if the hood and branch ducts are
part of the path of greatest resistance. As defined in Chapter 8 this is
the path from a hood to the fan that causes the most pressure loss. This
path governs the static pressure loss throughout the whole ventilating
system. The pressure loss through other hoods and branch ducts are

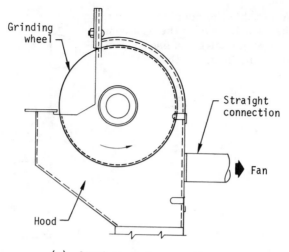

(a)   Straight duct connection
      at hood

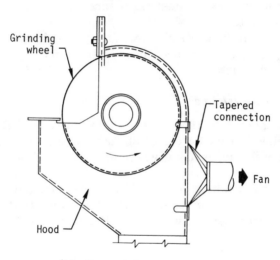

(b)   Tapered duct connection
      at hood

Figure 13.7   The tapered duct connection in (b)
increases airflow through the hood by about 7%
over the straight duct connection in (a).

adjusted by using dampers or small duct diameters that balance airflow
correctly. Reducing resistance in a hood or duct that is *not* part of the
path of greatest resistance will increase airflow through that branch but
will decrease or not change airflow through the other hoods. Reducing

Figure 13.8 Adjustable slot openings allow the slot width to be varied. Usually the best choice is the widest opening that gives good velocity distribution along the length of the slot.

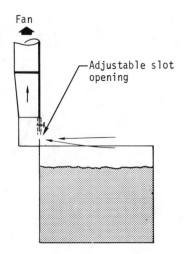

resistance in the path of greatest resistance will increase flow through all hoods in the system.

Since the main duct, air cleaner, fan, and exhaust stack are part of the path of greatest resistance, these are good places to look for ways to reduce airflow resistance.

Air Cleaner Resistance

The air cleaning device is the largest single source of airflow resistance in many systems. Except for electrostatic precipitators, a high contaminant removal efficiency is usually accompanied by a high pressure drop. If the existing air cleaner is replaced with an air cleaner with less airflow resistance, airflow will be increased. For filters, shortening the cleaning or replacement interval will reduce the resistance that the fan must work against during periods when the filter is dirty.[5]

Main Ducts

The main duct receives flow from branch ducts leading from each hood and carries the air to the fan. Lowering the pressure loss through this duct will increase airflow through the system. Chapters 5 and 8 discuss friction and turbulence as sources of pressure loss in ducts. Diameter enlargements and reductions should be designed to minimize pressure loss. Wide elbow curvature is better than tight elbows; elbows with turning vanes are better yet. Eliminating elbows is the best solution.

Fan Inlet and Outlet

The ultimate answer to insufficient flow in a ventilation system is the fan. Increasing the size or rotating speed of the fan will make up for many problems with the hoods, ducts, or air cleaners. Because these fan modifications are so important, they will be covered separately later in this chapter.

One fan item to check is the fan inlet connection from the duct to the fan. The fan wheel can do the greatest amount of work on incoming air only if the airflow is straight and uniform.[6] Spinning or nonuniform flow patterns reduce the fan's air volume and/or static pressure output. Major reasons for poor flow are elbows, dampers, duct entries, or other flow disturbances near the fan. When a new system is being designed, these interferences can be relocated before installation. Relocation is the best solution for an existing system, but installing flow straighteners in the inlet duct or turning vanes in elbows or fan inlet boxes will restore straight flow into the inlet (Figure 9.8).

Even with a straight inlet duct, spinning can occur in the duct. Cyclone-type air cleaners are noted for setting up corkscrew airflow. Use Pitot tube and temperature tests discussed in Chapter 12 to check for spinning or stratified inlet flow.

Fan rating curves and tables are developed from tests that use ideal fan inlet and outlet connections. If your system has poor connections, the fan will not perform as well as you might expect from data in the fan catalog.

Exhaust Stacks

Every system needs at least a short straight exhaust stack. A stack on the fan outlet helps to change high, uneven velocity patterns into a uniform flow. The result, believe it or not, is more efficient fan performance even though the stack is located after the fan in the ventilation system. This phenomenon is called *static regain* and occurs when velocity pressure is converted to static pressure in the stack. This static pressure helps to push the exhaust air out of the stack.

Static pressure regain is maximized by using a gradual enlargement or evasé on the discharge (Figure 9.12). The evasé reduces stack discharge velocity to a minimum. Even though low discharge velocity might be acceptable for a system with an air cleaner, a high stack velocity helps to disperse contaminants to avoid recirculating them through the building air intake system. However, increasing stack height after the evasé generally provides better dispersion of contaminants than high discharge velocity.

FAN MODIFICATIONS OR REPLACEMENT

There are four choices when you want to increase airflow through a ventilation system by changing an existing fan installation:

- Increase the rotating speed of the existing fan.
- Replace the existing fan with a larger or different type of fan.
- Separate the existing ventilation system into two or more smaller systems, each with its own fan.
- Add a second fan to the system either in series or in parallel with the existing fan.

Increasing Fan Speed

If you increase the rotating speed, the fan will deliver a greater volume of air against a higher static pressure. The increase in volume output is directly related to the increase in rotating speed; if you double the fan speed, the volume output doubles. Static pressure increases as the square of increases in fan speed; if you double the speed, the static pressure capability increases by a factor of 4. Unfortunately, moving more air against higher static pressure costs more money. The brake horsepower, which represents electrical power consumption, increases as the third power of increases in fan rotating speed; doubling the fan speed requires eight times more electrical power to run the fan ($2 \times 2 \times 2 = 8$). These and other fan laws were discussed in Chapter 9. Whenever you apply fan laws, remember that they all act together and are also related to the entire ventilation system. For example, increasing the airflow through an existing system increases the friction and turbulence in the ducts and, as a result, the fan will have to work against more resistance at the higher flow rate.

Example: An 18-in.-diameter ventilating system fan is moving 4000 ft³/min of air with a fan static pressure of 2.5 in. of water. A tachometer reading shows the rotating speed to be 1650 rev/min. The fan catalog lists 2.23 Brake Horsepower as the energy consumption at this operating point (Figure 13.9). Calculate the effect on speed, static pressure, and horsepower of increasing the airflow to 5000 ft³/min.

Answer: Since rotating speed and volume output are proportional, use Equation 9.3 to find the new rotating speed:

$$\frac{R_1}{R_2} = \frac{Q_1}{Q_2}$$

where R = fan rotating speed, rev/min
       Q = airflow, ft³/min

$$\frac{1650}{R_2} = \frac{4000}{5000}$$

$$R_2 = 2063 \text{ rev/min}$$

Calculate the new fan static pressure using Equation 9.4:

$$\left(\frac{R_1}{R_2}\right)^2 = \frac{FSP_1}{FSP_2}$$

CFM	2" SP		2½" SP		3" SP		3½" SP	
	RPM	BHP	RPM	BHP	RPM	BHP	RPM	BHP
1535								
1727								
1919								
2111	1276	0.94						
2303	1286	1.01						
2495	1298	1.06	1434	1.37				
2687	1315	1.14	1446	1.45	1567	1.77		
2879	1336	1.22	1461	1.54	1579	1.87	1694	2.23
3071	1361	1.31	1478	1.63	1593	1.97	1704	2.34
3263	1387	1.41	1499	1.73	1608	2.08	1717	2.45
3455	1415	1.51	1523	1.84	1628	2.19	1731	2.57
3647	1446	1.62	1549	1.96	1651	2.33	1750	2.71
3839	1480	1.74	1580	2.09	1677	2.46	1771	2.85
4031	1514	1.86	1610	2.23	1703	2.61	1794	3.01
4223	1552	2.01	1641	2.37	1732	2.76	1822	3.17
4415	1589	2.14	1677	2.53	1763	2.93	1849	3.34
4607	1628	2.31	1714	2.71	1797	3.11	1879	3.53
4799	1669	2.47	1752	2.88	1829	3.28	1910	3.73
4991	1710	2.64	1788	3.06	1867	3.51	1943	3.93
5375	1792	3.01	1869	3.47	1939	3.91	2012	4.37

CFM	4" SP		4½" SP		5" SP		5½" SP	
	RPM	BHP	RPM	BHP	RPM	BHP	RPM	BHP
3071	1809	2.72						
3263	1818	2.84	1920	3.25				
3455	1831	2.97	1930	3.39	2023	3.81	2115	4.26
3647	1848	3.11	1943	3.54	2034	3.97	2124	4.42
3839	1864	3.26	1958	3.69	2046	4.13	2134	4.59
4031	1886	3.43	1975	3.86	2062	4.31	2148	4.78
4223	1908	3.59	1993	4.03	2079	4.49	2163	4.97
4415	1932	3.77	2016	4.22	2098	4.69	2178	5.17
4607	1960	3.97	2039	4.42	2120	4.91	2197	5.38
4799	1987	4.17	2066	4.64	2142	5.11	2218	5.61
4991	2018	4.38	2094	4.86	2169	5.35	2242	5.85
5375	2083	4.85	2155	5.35	2225	5.85	2294	6.36
5759	2154	5.37	2219	5.87	2285	6.38	2354	6.95
6143	2227	5.92	2290	6.45	2352	6.99	2418	7.57
6527	2305	6.54	2365	7.09	2425	7.66	2484	8.23
6911	2385	7.21	2443	7.78	2500	8.37	2556	8.97
7295	2470	7.94	2522	8.52	2577	9.13	2631	9.75
7679	2552	8.71	2606	9.34	2658	9.98	2710	10.6

Figure 13.9 Fan rating table for an 18-inch-diameter backward inclined blade fan. CFM = Flow rate, ft³/min; RPM = Rotating speed, rev/min; SP = Fan static pressure, inches of water; BHP = Brake horsepower. (Source: Reference 7)

where FSP = fan static pressure, inches of water

$$\left(\frac{1650}{2063}\right)^2 = \frac{2.5}{FSP_2}$$

$$FSP_2 = 3.9 \text{ in. of water}$$

Calculate the new Brake Horsepower using Equation 9.6:

$$\left(\frac{R_1}{R_2}\right)^3 = \frac{BHP_1}{BHP_2}$$

where BHP = Brake Horsepower

$$\left(\frac{1650}{2063}\right)^3 = \frac{2.23}{BHP_2}$$

$$BHP_2 = 4.4 \text{ horsepower}$$

So a 5-horsepower motor is needed.

Whether or not the best solution lies in increasing the fan speed to achieve the desired performance is usually a straight economic decision. The cost of the added electrical power can be calculated over the projected life of the fan. For small increases in fan speed the existing fan motor may be adequate but the fan manufacturer's literature will show whether a higher horsepower motor is needed. Finally, every fan has a rated maximum safe speed based on structural strength. A quick check will show whether the existing fan is safe at the required speed.

Replacing the Fan

The fan must be replaced with a larger unit if the rotating speed calculated to give the required airflow or static pressure exceeds the maximum safe rotating speed in the fan literature.

Improving Efficiency

A second reason to replace, rather than speed up, the existing fan is to improve operating efficiency. Every fan has an operating range where efficiency is highest. If the airflow is too low, the fan blades "beat" the air rather than move it. This causes increased turbulence and noise, which consumes energy without adding to fan performance. Optimum efficiency occurs when the airflow over the fan blades is smooth. If the airflow is too great, the moving air breaks away from the fan blades and efficiency again drops. As an example, consider the 18-in.-diameter fan in the previous example. The fan rating table for that fan is shown in Figure 13.9. Using the previous example, with a fan output of 5000 ft³/min at 3.9 in. of water fan static pressure the fan is still operating within its range of maximum efficiency as indicated by the shaded band on the table. The shaded band represents operation within 3% of maximum efficiency. From Figure 13.9, the 18-in.-diameter fan needs 4.4 brake horsepower to move this air. Suppose a 15-in.-diameter fan were installed in the system. Figure 13.10 is the rating curve for this fan. Operating it at 5000 ft³/min and 3.9 in. of water fan static pressure would require about 6.8 brake horsepower, since the fan would be operating at lower efficiency. The 15-in. fan would therefore be too small for the job.

To decide between replacing an inadequate size fan or speeding it up despite the higher power costs, compare the replacement cost with

CFM	4" SP RPM	4" SP BHP	4½" SP RPM	4½" SP BHP	5" SP RPM	5" SP BHP	5½" SP RPM	5½" SP BHP
1934	2187	1.94	2308	2.23				
2063	2198	2.02	2318	2.32	2435	2.63	2551	2.97
2192	2214	2.11	2333	2.42	2445	2.73	2557	3.06
2321	2237	2.22	2349	2.52	2458	2.83	2566	3.17
2450	2261	2.34	2369	2.64	2475	2.95	2581	3.29
2579	2287	2.46	2394	2.77	2497	3.09	2600	3.43
2708	2319	2.61	2421	2.91	2519	3.22	2620	3.57
2837	2354	2.74	2451	3.05	2548	3.38	2644	3.72
2966	2389	2.91	2484	3.21	2577	3.54	2671	3.89
3095	2430	3.07	2520	3.39	2611	3.73	2699	4.06
3224	2470	3.24	2557	3.57	2646	3.91	2731	4.26
3353	2510	3.42	2599	3.77	2680	4.11	2766	4.46
3611	2601	3.83	2681	4.18	2761	4.54	2842	4.91
3869	2693	4.26	2773	4.65	2849	5.02	2925	5.41
4127	2792	4.74	2868	5.14	2940	5.54	3011	5.94
4385	2895	5.27	2964	5.67	3036	6.11	3103	6.52
4643	3000	5.83	3068	6.27	3133	6.71	3200	7.15
4901	3108	6.45	3174	6.91	3235	7.35	3298	7.81
5159	3216	7.09	3278	7.57	3342	8.07		

Figure 13.10 Fan rating table for a 15-inch-diameter backward inclined blade fan. CFM = Flow rate, ft³/min; RPM = Rotating speed, rev/min; SP = Fan static pressure, inches of water; BHP = Brake horsepower. (Source: Reference 7)

the cost of the additional power costs. Chapter 11 contains the equation for calculating the power costs of operating the two fans. Often the number of hours of operation per day or week govern whether it is more economical to replace a small fan.

Replacing a fan has two other advantages:

The range of maximum operating efficiency for a fan corresponds to the zone of lowest noise output. Speeding up a fan will always increase noise output, but the noise will be much greater if the fan is no longer operating within the range of maximum efficiency. If noise is a problem it may be worth additional expense to replace a small fan with one operating at peak efficiency.

Different types of fans have different efficiencies. As discussed in Chapter 9 there are three major types of centrifugal fans: radial blade, forward curved blade, and backward inclined blade. The airfoil fan is a special type of backward inclined blade fan. Radial blade fans are inefficient but necessary in dusty systems. Forward curved blade fans are less efficient than backward inclined blade units except at low airflows and static pressures. If the present fan is not a backward inclined blade fan, and dust or other factors do not preclude its use, check for cost savings from the use of a more efficient type of fan in the system.

Dividing the System

Dividing a multiple-hood ventilation system into two or more separate systems is a good idea if the resistance to airflow due to duct friction is excessive.[8] Duct friction and turbulent losses vary with the square of duct velocity; double the velocity and the duct losses increase by a factor of 4. The resistance to airflow at different airflows is illustrated in a system curve (Figure 13.11) prepared by calculating the pressure loss at several volumetric flow rates. The shape of the system curve reveals whether the ducts are adequate. If the curve is fairly flat between the present airflow volume and the desired volume (Curve A), then the ducts are adequate. If the system curve rises sharply (Curve B), then the ducts are too small for economical operation. This is because the added airflow causes a severe increase in pressure loss. Chapter 11 describes techniques for saving money by dividing inefficient systems into separate systems, each with its own fan.

Adding a Second Fan

If the ducts are adequate to handle the higher airflow without excessive pressure loss, then adding a second fan to the system may be feasible.

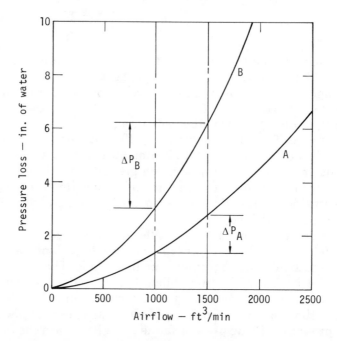

Figure 13.11   Pressure loss as a function of airflow for two similar ventilation systems. System A has ducts with larger diameters than system B.

The fans can be installed either in series or parallel (Figure 13.12). The most efficient way to use two fans in a single system is to install two identical fans so that each fan does equal work.[1] If you use two different size fans, it is possible for one of the fans to do all the work. Very often the cost of duct modifications reduces the cost savings of adding a second fan as compared to buying a larger single fan.

NOISE PROBLEMS

Noise from ventilating systems is rarely loud enough to be a hazard to hearing except occasionally in rooms that contain several fans, chillers, or other noisy equipment. Ventilation system noise can nevertheless be a nuisance, especially in normally quiet areas like laboratories or light assembly shops. The nuisance factor of a noise depends on its intensity, acoustic frequency distribution, and time duration as well as on the background noise from other sources.

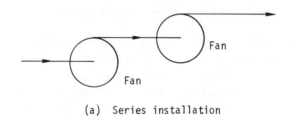

(a)   Series installation

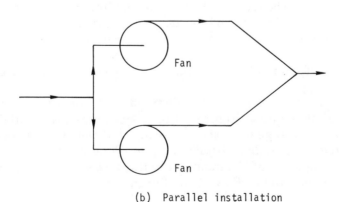

(b)   Parallel installation

Figure 13.12   Fan output can be increased by adding a second fan either in series or in parallel. The most efficient operation occurs when both fans are the same size. (Source: Reference 1)

Although most noise problems in the area exhausted by the system are caused by fan noise traveling through or along the ducts to the workroom, the moving air can also generate excessive noise in systems with high hood or duct velocities. As explained in Chapter 9, fan noise is caused by:

- Turbulence inside the fan.
- Mechanical noise from motors, bearings, and the fan belt or direct drive. Vibration is also considered mechanical noise.

Turbulence noise is related to the fan's operating efficiency. If the fan is operating within its zone of maximum efficiency and is rotating in the proper direction, there is not much you can do to reduce the amount of noise generated. You can, however, reduce the amount of noise reaching the work area.

Mechanical noise is usually not a source of annoyance in typical industrial systems.

Reducing Turbulence Noise Transmission

Turbulence noise travels up and down the ducts from the fan. If there is only a short duct length between the fan and a hood, the noise problem will be magnified. One solution, though not widely used in industrial exhaust systems, is duct acoustical treatment. Acoustical treatment works best when the sound-absorbing material is placed perpendicular to the direction of noise travel. For this reason, acoustical lining in duct elbows is a good way to reduce noise. The amount of material needed depends on the sound absorption characteristics of the material and the amount that the noise is to be reduced. Installing duct lining in an existing elbow may restrict flow because of the decrease in duct diameter; a larger diameter duct elbow may therefore be needed.[9] Installing acoustical lining on turning vanes in an elbow is even more effective, since a larger surface area of sound-absorbing material is exposed to the noise (Figure 13.13).

Silencers, which are devices mounted in the duct to reduce sound, are another possibility (Figure 13.14). Some fans are available with silencers as a package unit, thus assuring that the fan and silencer are aerodynamically matched.[10] If a fan and silencer package is not available, the silencer should be mounted in the duct some distance from the fan to avoid spinning airflow at the fan inlet.

Reducing Mechanical Noise Transmission

Mechanical noise is caused by moving or vibrating fan components. It can travel to the work area by conduction through the building structure

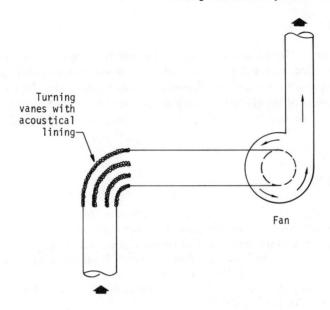

Figure 13.13   Acoustical lining in ducts or in turning
vanes reduces noise transmission along the air path.

or along the ventilation system ducts. Popular ways to reduce this noise
are:

1.   Isolating the fan from the building and ducts as shown in Figure
      9.23.
2.   Using an acoustical enclosure around the fan, motor, and drive if
      noise radiating from the fan housing or motor is a problem.

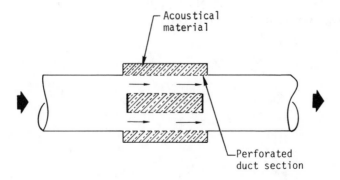

Figure 13.14   Silencers mounted in the duct or as part
of the fan reduce noise transmission to the work area.
(Source: Reference 11)

SUMMARY

This chapter outlined some of the common problems with ventilation systems for contaminant control and some solutions to these problems. The primary problems are insufficient airflow through hoods, excessively high contaminant levels in the work areas, and noise complaints. Careful investigation is needed to diagnose the problems and find the best solutions.

REFERENCES

1. ACGIH Committee on Industrial Ventilation. *Industrial Ventilation—A Manual of Recommended Practice,* 17th Ed. (Lansing, Michigan: American Conference of Governmental Industrial Hygienists, 1982).
2. Trickler, C.J. "Field Testing of Fan Systems," Engineering Letter E-3 (Chicago, Illinois: The New York Blower Company).
3. Hemeon, W.C.L. *Plant and Process Ventilation* (New York, New York: Industrial Press, Inc., 1963).
4. National Safety Council. "Checking Performance of Local Exhaust Systems," Data Sheet 438 (Chicago, Illinois: National Safety Council, 1963).
5. Burchsted, C.A. and A.B. Fuller. "Design, Construction and Testing of High-Efficiency Air Filtration Systems for Nuclear Applications," U.S. Atomic Energy Commission, Report ORNL–NSIC–65 (Oak Ridge, Tennessee: U.S. AEC Divisional of Technical Information, 1970).
6. Trickler, C.J. "Effect of System Design on the Fan," Engineering Letter E-4 (Chicago, Illinois: The New York Blower Company).
7. Catalog, New York Blower Company, Chicago, Illinois.
8. Trickler, C.J. "How to Get More Air From Your Fan," Engineering Letter A-5 (Chicago, Illinois: The New York Blower Company).
9. Trickler, C.J. "How to Select Centrifugal Fans for Quiet Operation," Engineering Letter E-13 (Chicago, Illinois: The New York Blower Company).
10. Trickler, C.J. "New Concept: Fan/Silencer Package," *Air Conditioning, Heating and Ventilation* 62, No. 12 (1965).
11. Catalog, Aerovent, Inc., Piqua, Ohio.

APPENDIX A

Symbols and Abbreviations

SYMBOLS

Symbols	Description	Typical Units
A	Area	ft^2
A	Air exchange rate	air changes/hr
a	Experimental coefficient for Eq (7.1)	—
B	Constant to adjust LEL for high temperatures	—
b	Experimental coefficient for Eq. (7.1)	—
BHP	Brake horsepower	horsepower
C	Airborne concentration of contaminant	ppm, mg/m^3
C	Safety factor for dilution ventilation for fire (Chapters 1 and 10)	—
C_e	Hood coefficient of entry	—
CF	Cost of fuel	$/lb
D	Duct diameter	in., ft
DD	Number of degree-days	—
E	Equivalent exposure to a mixture of contaminants	—
E	Motor efficiency (Chapter 12)	—
F	Conversion factor (Chapter 1)	—
f	Friction loss coefficient	—
F_f	Friction loss	ft of air
F_h	Hood entry loss factor	—
F_s	Safety factor for poor air distribution into enclosure	—
FSP	Fan Static Pressure	in. of water
g	Gravitational acceleration	ft/sec^2
H	Total sensible heat load (Chapter 10)	Btu/hr
H	Heat required (Chapter 11)	Btu/hr
H	Velocity head (Chapter 5)	ft of air
h_f	Duct friction loss	in. of water
h'_f	Duct friction loss	VP equiv.
h_t	Turbulent pressure loss	in. of water
K	Safety factor for dilution ventilation for health	—
L	Acceptable airborne contaminant level (Chapters 1 and 3)	ppm, mg/m^3
L	Length	ft
LEL	Lower Explosive Limit	percent
M	Molecular weight	—
ME	Mechanical Efficiency	—

Symbols	*Description*	*Typical Units*
P	Perimeter of hood opening	ft
PF	Motor power factor	—
Q	Volumetric airflow rate	ft^3/min
q	Available heat per unit of fuel	Btu/lb
R	Fan rotating speed	rev/min
RPM	Respirable Particulate Matter	—
SP	Static Pressure	in. of water
|SP|	Static Pressure (absolute value)	in. of water
SP_h	Hood static pressure	in. of water
sp. gr.	Specific gravity of liquid (water = 1.0)	—
T	Temperature (Chapter 1)	°F
T	Time	hr
T	Operating time (Chapter 11)	hr/week
t_{bp}	Contaminant boiling point	°C
t_d	Room air temperature	°F
t_{10}	Time until 10% breakthrough of activated carbon occurs	min
Δt	Temperature increase	°F
Δt	Elapsed time of test (Chapter 2)	hr
TLV	Threshold Limit Value	ppm, mg/m^3
TP	Total Pressure	in. of water
TWA	Time Weighted Average exposure	ppm, mg/m^3
V	Velocity	ft/min
v	Velocity	ft/sec
Vol	Room volume	ft^3/air change
VP	Velocity Pressure	in. of water
W	Quantity of solvent used per time interval	pints/hr
W_c	Weight of activated carbon	lb
X	Distance away from hood opening	ft

actual	at actual conditions
adjusted	corrected for ambient conditions
avg	average
calc	as calculated
d	duct
design	from original design calculations
dn	downstream
equiv	equivalent
f	friction
h	hood
inlet	fan inlet
max	maximum theoretical
outlet	fan outlet
t	turbulent
up	upstream
x	at x distance from hood opening

ABBREVIATIONS

Abbrevia-tion	*Description*
Btu	British thermal unit
ft	feet
hr	hour
in.	inches
μm	micrometers
mg/m^3	milligrams per cubic meter
min	minute
ppm	parts per million by volume
psi	pounds per square inch
psia	pounds per square inch absolute
psig	pounds per square inch gauge
rev	revolutions

Glossary

Absorption: diffusion process in which molecules are transferred from the gas phase to a liquid.

Acceleration loss: the energy required to accelerate air to a higher velocity.

Administrative controls: methods of controlling employee exposures to contaminants by job rotation, work assignment, or time periods away from the contaminant.

Adsorption: process in which a gas or vapor adheres to the surface of a solid (usually a porous material).

Air cleaner: a device that removes airborne contaminants from exhaust air.

Balancing by dampers: method for designing local exhaust system ducts using adjustable dampers to distribute airflow after installation.

Balancing by static pressure: method for designing local exhaust system ducts by selecting the duct diameters that generate the static pressure to distribute airflow without dampers.

Blast gates: adjustable sliding dampers used to distribute airflow.

Brake horsepower: the power to spin the fan neglecting losses in the drive connecting the fan and motor. One horsepower equals 33,000 ft-lb per min.

Brake horsepower curve: a graphical representation of brake horsepower at different airflow rates for a fan.

Branch duct entry: the point where a branch or secondary duct joins a main duct.

Branch (or path) of greatest resistance: the path from a hood to the fan and exhaust stack in a ventilation system that causes the most pressure loss.

Breathing zone sample: an air sample collected in the breathing area of a worker to assess inhalation exposure to airborne contaminants.

Carcinogen: a substance that causes cancer.

Ceiling exposure limit: an OSHA standard, setting the maximum concentration of a contaminant to which a worker may be exposed.

Coefficient of entry (C_e): the actual airflow rate into a hood compared with the theoretical maximum airflow with the same hood static pressure.

Comfort ventilation: airflow intended to remove heat, odor, and cigarette smoke.

Dampers: adjustable sources of airflow resistance used to distribute airflow in a ventilation system.

Degree day: a unit used to estimate heating and cooling costs. For example, on a day when the mean temperature is less than 65°F, there is the same number of degree days as the mean temperature is below 65°F.

Differential pressure: the difference in static pressure between two locations.

Duct: a conduit used for conveying air at low pressures.

Dust: small solid particles formed by the breaking up of larger particles.

Elbow curvature: the ratio of the radius of curvature of a duct elbow to the duct diameter.

Engineering controls: methods of controlling employee exposures by modifying the source or reducing the quantity of contaminants released into the workroom environment.

Equivalent exposure: the employee's exposure to a mixture of airborne contaminants calculated by summing the fraction of the allowable exposure to each component in the mixture.

Evasé: a section of duct in the exhaust stack that gradually increases in diameter.

Fan, airfoil: a type of backward inclined blade fan with blades that have an airfoil cross-section.

Fan, axial: a fan in which airflow is parallel to the fan shaft and movement is induced by a screw-like action.

Fan, backward inclined blade: a centrifugal fan with blades that are inclined in a direction opposite to fan rotation.

Fan, centrifugal: a fan in which the air leaves the fan in a direction perpendicular to the entry direction.

Fan, forward curved blade: a centrifugal fan with blades that are inclined in the direction of fan rotation.

Fan, propeller: an axial fan employing a propeller to move air.

Fan, radial blade: a centrifugal fan with blades extending out radially from the fan wheel shaft.

Fan, tube axial: an axial fan mounted in a duct section.

Fan, vane axial: an axial flow fan mounted in a duct section with vanes to straighten the airflow and increase pressure.

Fan inlet vanes or dampers: devices installed at the fan inlet to impart spin to air entering the fan to change fan performance.

Fan laws: statements and equations that describe the relationship between fan volume, pressure, brake horsepower, size and rotating speed.

Fan rating curve or table: data that describe the volumetric output of a fan at different static pressures.

Fan static pressure: the pressure added to the system by the fan. It equals the sum of pressure losses in the system minus the velocity pressure in the air at the fan inlet.

Filter, absolute: see **Filter, HEPA.**

Filter, HEPA: high-efficiency particulate air filter that is at least 99.97% efficient in removing thermally generated monodisperse dioctylphthalate smoke particles with a diameter of 0.3 μm.

Flange: a rim or edge added to a hood to reduce the quantity of air entering the hood from behind the hood.

Friction loss: the pressure loss due to friction.

Fumes: small solid particles formed by the condensation of vapors of solid materials.

Gas: formless fluids which tend to occupy an entire space uniformly at ordinary temperatures.

Glove box: a sealed enclosure in which all handling of items inside the box is carried out through long impervious gloves sealed to ports in the walls of the enclosure.

Grab sample: an air sample collected over a short time.

Hood: a location where air enters the ventilation system to control contaminants.

Hood, canopy: a hood that is located over a source of emissions.

Hood, capturing: a hood with sufficient airflow to reach outside of the hood to draw in contaminants.

Hood, enclosing: a hood that encloses the contaminant source.

Hood, lateral: see Hood, capturing.

Hood, receiving: a hood sized and located to catch a stream of contaminants or contaminated air directed at the hood.

Hood, slot: a hood consisting of a narrow slot leading into a plenum chamber under suction to distribute air velocity along the length of the slot.

Hood entry loss: the pressure loss from turbulence and friction as air enters the ventilation system.

Hood static pressure: the suction or static pressure in a duct near a hood. It represents the suction that is available to draw air into the hood.

Humidity, absolute: the weight of water vapor per unit volume of air.

Humidity, relative: the ratio of the mole fraction of water vapor present in the air to the mole fraction present in saturated air at the same temperature and pressure.

Humidity, specific: the weight of water vapor per unit weight of dry air.

Inclined manometer: a manometer that amplifies the vertical movement of a water column through the use of an inclined leg.

Lower Explosive Limit (LEL): the lower limit of flammability or explosibility of a gas or vapor at ordinary ambient temperatures expressed as percent of gas or vapor in the air by volume.

Make-up air: air introduced into an area being exhausted to replace the air that is removed.

Manometer: a device for measuring pressure. Usually it is a U-shaped tube partially filled with liquid, constructed so that the amount of displacement of the liquid indicates the pressure exerted on the instrument.

Mechanical efficiency curve: a graphical representation of a fan's relative efficiency in moving air at different airflow rates and static pressures.

Mist: small droplets of materials that are ordinarily liquids at normal temperature and pressure.

Noise, mechanical: noise from friction or vibration in the machinery rather than due to moving air. Bearing noise and drive noise are typical examples.

Noise, turbulent: noise caused by the air moving through the fan and system.

Null point: the distance from a contaminant source that the initial energy or velocity of the contaminants is dissipated, and the material can be captured by a hood.

Operating point: the volumetric airflow and static pressure that a fan is producing.

Partial ventilation system: a local exhaust system designed with airflow less than that required for all hoods in the system. Airflow to hoods not in use is shut off so there is sufficient airflow through hoods being used.

Particulate: airborne material that has a relatively fixed shape and volume such as dusts, mists, smokes, and fumes.

Peak above ceiling exposure limit: an OSHA exposure standard that permits a short-term peak exposure above the ceiling limit.

Personal protective equipment: devices worn by the worker to protect against hazards in the environment. Respirators, gloves, and ear protectors are examples.

Pitot tube: a device for measuring pressure consisting of two concentric tubes arranged to measure total and static pressures.

Plenum chamber: an air compartment connected to one or more ducts or connected to a slot in a hood used for air distribution.

Pressure: the normal force exerted by a homogeneous liquid or gas, per unit of area, on the walls of its container.

Pressure, static: the potential pressure exerted in all directions by a fluid at rest.

Pressure, total: the algebraic sum of velocity pressure and static pressure (with regard to sign).

Pressure, velocity: the kinetic pressure exerted in the direction of flow necessary to cause a fluid at rest to flow at a given velocity.

Pressure drop: the difference in static pressure measured at two locations in a ventilation system due to friction or turbulence.

Pressure loss: energy lost from a ventilation system through friction or turbulence.

Psychometric chart: a graphical representation of the thermodynamic properties of moist air.

Push-pull hood: a hood consisting of an air supply system on one side of the contaminant source blowing across the source and into an exhaust hood on the other side.

Respirable size particulates: particulates in the size range that permit them to penetrate to the lungs upon inhalation.

Smoke: an aerosol, usually, but not necessarily, solid formed by combustion or sublimation.

Standard air density: the density of air, 0.075 lb/ft^3, at standard conditions.

Standard conditions: in industrial ventilation, 70°F, 50% relative humidity, and 29.92 in. of mercury atmospheric pressure.

Static pressure curve: a graphical representation of the volumetric output and fan static pressure relationship for a fan operating at a specific rotating speed.

Static pressure regain: the increase in static pressure in a system as air velocity decreases and velocity pressure is converted into static pressure according to Bernoulli's Theorem.

System curve: a graphical representation of the static pressure (resistance) in a ventilation system at different airflow rates.

Tachometer: a device for measuring rotating speed.

Temperature, dry-bulb: air temperature, measured with a thermometer, after correction for radiation.

Temperature, wet-bulb: the temperature at which liquid or solid water, by evaporating into air, can bring the air to saturation adiabatically at the same temperature.

Tempering: the process of heating or cooling make-up air to the proper temperature.

Threshold Limit Value (TLV):   values for airborne concentrations of chemical substances and physical stresses for evaluating occupational exposures, published by the American Conference of Governmental Industrial Hygienists.

Time-weighted average (TWA) exposure: an employee's average exposure to a contaminant over the work period or shift.

Ton of air conditioning: a useful measure of refrigerating rate equal to 12,000 Btu/hr.

Transport velocity: the duct velocity needed to prevent settling of airborne dusts or other particulates.

Turbulence loss: the pressure or energy lost from a ventilation system through air turbulence.

Turning vanes: curved pieces added to elbows or fan inlet boxes to direct air and so reduce turbulence losses.

Vapor: the gaseous form of substances that are normally solid or liquid at ambient temperatures.

Velocity, capture: the air velocity needed to draw contaminants into the hood.

Velocity, face: the inward air velocity in the plane of openings into an enclosure.

Velometer: a device for measuring air velocity.

Vena contracta: the reduction in the diameter of a flowing airstream at hood entries and other locations.

Ventilation, dilution: airflow designed to dilute contaminants to acceptable levels.

Ventilation, general: see **Ventilation, dilution.**

Ventilation, local exhaust: a system, usually consisting of hoods, ducts, air cleaner, and fan, that captures or contains contaminants at their source for removal from the work environment.

Ventilation, mechanical: air movement caused by a fan or other air moving device.

Ventilation, natural: air movement caused by wind, temperature difference, or other nonmechanical factors.

APPENDIX C

Heating and Cooling Degree Days

Annual Normals*

	Heating	*Cooling*
ALABAMA		
Auburn	2528	2008
Huntsville	3302	1808
Mobile	1684	2577
Tuscaloosa	2626	2138
ALASKA		
Anchorage	10911	0
Circle Hot Springs	15671	17
Juneau	9007	0
Nome	14325	0
ARIZONA		
Flagstaff	7322	140
Holbrook	4826	1091
Phoenix	1552	3508
Tombstone	2350	1897
ARKANSAS		
Blytheville	3385	1982
Fayetteville	3839	1487
Little Rock	3354	1925
Magnolia	2468	2179
CALIFORNIA		
Bakersfield	2185	2179
Los Angeles	1819	615
Sacramento	2843	1159
San Jose	2416	444
COLORADO		
Denver	5505	742
Durango	6930	188
Grand Junction	5605	1140
Pueblo	5394	981
CONNECTICUT		
Hartford	6350	584
New Haven	5793	573

	Heating	*Cooling*
DELAWARE		
Bridgeville	4495	1053
Dover	4357	1166
Wilmington	4940	992
FLORIDA		
Jacksonville	1327	2596
Miami	206	4038
Panama City	1388	2778
Tampa	718	3366
GEORGIA		
Athens	2975	1722
Atlanta	3095	1589
Macon	2240	2294
Savannah	1952	2317
HAWAII		
Hilo	0	3066
Honolulu	0	4221
Lihue	0	3719
IDAHO		
Boise	5833	714
Bonners Ferry	7091	212
Idaho Falls	7888	286
Lewiston	5464	657
ILLINOIS		
Chicago	6497	664
Moline	6395	893
Peoria	6098	968
Sparta	4388	1506
INDIANA		
Bloomington	4905	1177
Fort Wayne	6209	748
Indianapolis	5577	974
South Bend	6462	695
IOWA		
Burlington	6149	994
Des Moines	6710	928
Dubuque	7277	606
Sioux City	6953	932
KANSAS		
Atchison	5197	1327
Dodge City	5046	1411
Salina	4992	1627
Wichita	4687	1673
KENTUCKY		
Ashland	4555	1173
Bowling Green	4219	1467
Louisville	4640	1268
Paducah	4025	1592
LOUISIANA		
Baton Rouge	1670	2585
Lake Charles	1498	2739
New Orleans	1465	2706
Shreveport	2167	2538

	Heating	*Cooling*
MAINE		
Eastport	7933	30
Greenville	9387	109
Portland	7498	252
MARYLAND		
Baltimore	4729	1108
Easton	4299	1195
Frederick	5059	948
Salisbury	3985	1220
MASSACHUSETTS		
Boston	5621	661
Springfield	5844	740
Worcester	6848	387
MICHIGAN		
Detroit	6228	743
Marquette	8351	216
Midland	6750	569
Sault Ste. Marie	9193	139
MINNESOTA		
Duluth	9757	176
International Falls	10547	176
Minneapolis–		
St. Paul	8310	527
Winnebago	8208	588
MISSISSIPPI		
Biloxi City	1496	2682
Hattiesburg	1921	2393
Natchez	1921	2475
Yazoo City	2419	2307
MISSOURI		
Jackson	4175	1545
Lebanon	4546	1334
St. Joseph	5440	1334
St. Louis	4486	1640
MONTANA		
Billings	7265	498
Glendive	7774	729
Helena	8190	256
Lewistown	8586	192
NEBRASKA		
Atkinson	6825	842
North Platte	6743	802
Omaha	6049	1173
Oshkosh	6501	725
NEVADA		
Carson City	5753	354
Elko	7483	342
Elly	7814	207
Las Vegas	2601	2946
NEW HAMPSHIRE		
Manchester	7101	378

	Heating	*Cooling*
NEW JERSEY		
Glassboro	4894	996
Newark	5034	1024
New Brunswick	5111	842
Trenton	4947	968
NEW MEXICO		
Albuquerque	4292	1316
Deming	3294	1687
Hobbs	2985	1828
Santa Fe	6007	417
NEW YORK		
Buffalo	6927	437
New York	4848	1068
Poughkeepsie	5824	809
Schenectady	6817	642
NORTH CAROLINA		
Asheville	4237	872
Durham	3439	1452
Greenville	3060	1697
Wilmington	2433	1964
NORTH DAKOTA		
Bismarck	9044	487
Grand Forks	9876	384
Minot	9407	370
OHIO		
Cincinnati	5070	1080
Cleveland	6154	613
Columbus	5702	809
Toledo	6381	685
OKLAHOMA		
Bartlesville	3876	1838
Oklahoma City	3695	1876
Stillwater	3631	1947
Tulsa	3680	1949
OREGON		
Klamath Falls	6516	286
Portland	4792	300
Redmond	6411	149
PENNSYLVANIA		
Phoenixville	5114	950
Pittsburgh	5930	647
Scranton	6114	630
State College	6132	583
RHODE ISLAND		
Providence	5972	532
SOUTH CAROLINA		
Charleston	1904	2354
Columbia	2598	2087
Greenville	3095	1650
Marion	2487	1912

	Heating	Cooling
SOUTH DAKOTA		
Eureka	8727	513
Pierre	7677	858
Rapid City	7324	661
Watertown	8796	477
TENNESSEE		
Chattanooga	3505	1636
Knoxville	3478	1569
Memphis	3227	2029
Nashville	3696	1694
TEXAS		
Dallas	2290	2755
El Paso	2678	2098
Houston	1434	2889
San Antonio	1570	2994
UTAH		
Cedar City	6137	615
Moab	4664	1521
Ogden	5872	946
Salt Lake City	5983	927
VERMONT		
Burlington	7876	396
Rutland	7172	346
VIRGINIA		
Charlottesville	4162	1263
Norfolk	3488	1441
Richmond	3939	1353
Roanoke	4307	1030
WASHINGTON		
Seattle	5185	129
Spokane	6835	388
Walla Walla	4835	862
Yakima	6009	479
WEST VIRGINIA		
Beckley	5615	490
Elkins	5975	389
Huntington	4624	1098
Wheeling	5402	851
WISCONSIN		
Kewaunee	7710	286
La Crosse	7417	695
Milwaukee	7444	450
Wausau	8586	355
WYOMING		
Cheyenne	7255	327
Green River	8137	253
Sheridan	7708	446
Sundance	8151	381

*Base for degree-days is 65°F.
Source: U.S. Department of Commerce. "Monthly Normals of Temperature, Precipitation, and Heating and Cooling Degree Days 1941–70," Climatography of the U.S. No. 81 (Asheville, North Carolina: National Climatic Center, 1973).

Index

VELOCITY PRESSURE CALCULATION SHEET

1	Branch or Main Duct No.								
2	Air volume	ft^3/min							
3	COMPLETE FOR SLOT HOODS ONLY:								
	a	Slot area	ft^2						
	b	Slot velocity (Line 2 ÷ Line 3a)	ft/min						
	c	VP_{slot} (Table 8-4)	in. H_2O/VP_{slot}						
4	Duct length	ft							
5	Duct diameter	in.							
6	Duct area (Table 8-3)	ft^2							
7	Duct velocity (Line 2 ÷ Line 6)	ft/min							
8	VP_{duct} (Table 8-4)	in. H_2O/VP_d							
9	Straight duct friction factor (Fig 8-1)	$VP_d/100ft$							
10	Elbow radius	R/d							
11	Elbow pressure loss factor (Fig. 8-2)	$VP_d/elbow$							

PART A. LOSSES CALCULATED IN UNITS OF VP_{duct}

12	Acceleration loss	$1.0\ VP_d$							
13	Hood entry loss (Table 8-2, etc.)	VP_d							
14	Straight duct friction loss (Lines 4 × 9) ÷ 100								
15	Elbow loss (No. elbows × Line 11)	VP_d							
16	Branch duct entry loss (Figure 8-3)	VP_d							
17	Other								
18	Subtotal (Part A)	VP_d							